SOUND EFFECTS

SOUND EFFECTS

Radio, TV, and Film

ROBERT L. MOTT

FOCAL PRESS
Boston   London

Focal Press is an imprint of Butterworth Publishers.

Cover illustration from a photograph by Cinda Yank Mott. Used with permission.

Library of Congress Cataloging-in-Publication Data
Mott, Robert L.
    Sound effects : radio, TV, and film / Robert L. Mott.
        p.   cm.
    Includes bibliographical references.
    ISBN 0-240-80029-X
    1. Sound—Recording and reproducing.   2. Sound studios.   3. Radio
broadcasting—Sound effects.   4. Motion pictures—Sound effects.
5. Television—Sound effects.   I. Title.
TK7881.4.M67   1989
778.5'344—dc20   89-36367
                                                                   CIP

British Library Cataloguing in Publication Data
Mott, Robert L.
    Sound effects.
    1. Sound effects
    I. Title
    621.389'3
    ISBN 0-240-80029-X

Butterworth Publishers
80 Montvale Avenue
Stoneham, MA 02180

10 9 8 7 6 5 4 3 2 1

Printed in the United States of America

In fond memory of Professor Paul Courtland Smith, University of San Francisco, Radio and Television Department, who kept insisting that I write this book, and to my wife, Cinda, who saw to it that it was completed.

CONTENTS

CHAPTER 6 ——————————————————————
CREATING SOUND EFFECTS   115

CHAPTER 7 ——————————————————————
RADIO SOUND EFFECTS   139

CHAPTER 8 ——————————————————————
SOUND EFFECTS FOR TELEVISION   156

CHAPTER 9 ——————————————————————
SOUND EFFECTS FOR FILMS   181

CHAPTER 10 ——————————————————————
FOLEYING   192

PREFACE

The art of sound effects began when radio and film were first struggling for recognition. How sound effects served radio is legendary; far less is known of its role in film and television.

Since its inception, the creation of sounds has always been an art form that required technical knowledge. However, I found that although there is an abundance of excellent books written on the scientific aspects of sounds, there are virtually no books that comprehensively deal with the art of sound effects; I felt this was the area where my book would be of most value.

Having worked in sound effects since the days of live radio and television, I have seen many innovative and exciting changes in the technical manner in which sounds are created. Although this technology has provided a dramatic improvement in the actual sound of the effects, the reason for doing sound effects has changed very little. That is the focus of this book: to make you more aware of sounds and their more effective use, regardless of how they are produced.

If you still think of sound effects as limited to things—doorbells, dog barks, or crinkling cellophane to make the sound of fire—I think you are in for a pleasant surprise.

Because so little information is available on the subject of sound effects, I have attempted to make this book as comprehensive as possible. In order to do this, I have enlisted the views of other professionals in the media fields. Although many of the views in this book are those shared by other sound effects artists, I have purposely sought the insights and advice of other professionals. Some are engineers, while others are writers, actors, directors, and producers—all professionals who understand and appreciate the contributions sound effects can make to a story.

This book, therefore, is a collaborative effort to give the reader a rare opportunity to learn how the use of sound effects has evolved from the early days of radio to the art form it is today.

Following are some of those people that made this book possible. A warm and grateful acknowledgment to each of them.

Robert L. Mott

ACKNOWLEDGMENTS

George Balzaar—Television comedy writer
Howard Beal—Sound editor, Paramount Pictures
Mel Blanc—Vocal effects artist, Universal Studios
Bill Brown—CBS sound effects artist
Mike Caruso—NBC camera operator
Clark Casey—CBS sound effects artist
Ken Corday—Executive Producer, *Days of Our Lives*
Keene Crockett—ABC sound effects artist
Dane Davis—Sound designer
Ray Erlenborn—CBS sound effects artist
Monty Fraser—NBC sound effects artist
Walt Guftason—ABC sound effects artist
Paul Hockman—MGM sound editor
Robert Holmes—President, Communication Group Ltd.
Karen Kearns—Professor of Radio and Television, University of California, Northridge
Ray Kemper—Sound effects artist; writer
Wes Kenny—Executive Producer, *General Hospital*
Bill Levitsky—NBC sound mixer
Tom McLoughlin—Writer; director
Gerry McCarty—CBS sound effects artist
Shawn Murphy—Production and music mixer
Harry Nelson—ABC/NBC sound effects artist
Sam Patterson—NBC Post-production editor
Virgil Reimer—NBC sound effects artist; director
Paul Smith—Professor of Radio and Television, University of San Francisco
Bill Stinson—Sound/music editor, Paramount Pictures

SOUND EFFECTS

A HISTORY OF SOUND EFFECTS

THE BEGINNING OF RADIO DRAMA

Radio drama has been accurately described as "the theater of the mind," and the success or failure of any given program was in direct relationship to its ability to involve the radio listener's imagination.

The listening audience gave little thought to the ages or appearances of those early radio actors. The important concern was, what did their voice make them sound like?

It was for this reason that balding, middle-aged men could portray young, virile Adonises, while women very often played the parts written for both young girls *and* boys. But the epitome of adaptability was reached on the old radio show *Beulah*. On this highly rated show, the starring role of the black maid was successfully played by a white man!

If radio was the "theatre of the mind," it most certainly was also "the drama of deceit." And yet, as magical as all those versatile voices were, there was an element lacking that radio desperately needed if it was going to make its dramatic shows more realistic and exciting. It was one thing to tell a story using clever dialogue, spoken by talented actors, but unless the audience could visualize what these disembodied voices were actually doing and where the scenes were taking place, it soon became very confusing (Figure 1-1).

In those early days, some critics compared listening to a dramatic show on radio with attending the theater blindfolded. Although this was a rather extreme commentary on an industry so young, there was enough truth in it to make the champions of radio wince. And yet, radio was asking a great

FIGURE 1–1 Networks, sponsors, and the performers themselves were extremely sensitive about how the listening public perceived their radio characters. To preserve this image, many programs sent out only those publicity photographs showing the stars in the makeup and costumes of the characters they played. That is why this candid shot of these two nattily dressed tennis players is so rare. On the left is Freeman Gosden, and on the right is Charles Correll—better known to millions of listeners for over a quarter of a century as the lovable black comedians, "Amos 'n' Andy."

deal from its listeners. Unlike the familiar theater, audiences were without the enormous benefit of seeing what the actors were doing. With radio, all the action had to be implied with the actors' dialogue.

Some Growing Pains

Training people to work in this new and mysterious medium was slow and arduous. For that reason, radio began drawing upon the experienced people in the theater for much of their talent. Although this seemed like an excellent solution, it actually created some new problems. Actors who spent years honing their skills to be able to convey emotions by a slump of the shoulder or an arched eyebrow found these physical movements of little value to an unseeing audience. Furthermore, the rich, resonant voices that could be heard in the last row of the balcony now had to be retrained to comply with the demands of a mysterious electronic gadget called a microphone.

Even the writers from the theater were finding it difficult to make the transition. They too were accustomed to the visual contributions the actors made to prevent the monotony of putting everything into words. And just how long would radio audiences keep accepting such dialogue as, "Isn't that John's car I just heard drive up?" Especially when they never heard the sound of John's car. Yes, this was indeed an extremely painful growing period for radio.

Radio's Unique Problems

When actors in the theater are late with their entrances or are slow in delivering lines, it is called a "stage wait." Although these moments are awkward, they are hardly a reason for the audience to leave the theater en masse in search of another play. Yet this was precisely the case with radio's listening audience.

Because of the newness of the medium, many of the technical problems had yet to be worked out. This was true both in the actual transmission of the programs and in the questionable reliability of the home receivers. Even the atmospheric conditions had an influence on how well you received certain programs. During thunderstorms, most stations were either silent or transmitted such loud, annoying static that hearing your favorite program was impossible.

As a result of these and other transmission and reception problems, any silence that extended beyond a few seconds caused an anxious listener to dial to another station to see if this expensive new gadget was still operating properly. After all, this was the era of the Great Depression, and money didn't grow on trees!

Even today, when your television set loses its audio, see how quickly the station flashes a "please stand by" card. This is to assure you that there is nothing wrong with your receiver and to dissuade you from turning to another channel. But what could radio do when it lost its sound?

One thing radio station owners and producers didn't want was to have their valued listeners tuning to competing stations to see if it was the fault of their radio or the station they were listening to. After all, prospective advertisers were being enticed away from the printed media by promises of a guaranteed listening audience . . . and they better deliver!

Therefore, silence was to be avoided at all costs. And much to the distress of the already beleaguered actors, directors, and writers, this included any extended dramatic pauses that were so much of a part of everyone's training in the theater. Now, in radio, these tools were referred to unceremoniously as *"dead air!"*

This fear of silence was so acute that worried station owners paid a house musician to be on constant emergency studio standby. If there was any delay or silence that extended beyond what could be logically justified, organ music filled the airways to assuage the home listeners and to assure them that everything was just fine in radioland.

THE DISCOVERY OF SOUND EFFECTS

Radio's preoccupation with a constant "something" being on the air was becoming annoying to an audience that was accustomed to forms of entertainment where they knew exactly what was going on at all times. The novelty of hearing music and voices "magically without any wires" was beginning to pall. Now they wanted to be entertained.

As radio scurried desperately about to fill this need, it made a rather obvious but nevertheless startling discovery. Up until now radio had been so preoccupied with imitating the more familiar forms of entertainment found in the theater that it hadn't had time to realize what radio had to offer was unique.

Unlike the theater, radio didn't need expensive and elaborate costumes. It didn't need makeup or props. It didn't even need scenery or exotic locations. Radio had something better . . . the listener's imagination!

All radio had to do to avail itself of this magical world of fantasy was to create images through the suggestion of sound. After all, if a white man could convincingly portray the part of a black woman simply by the way he made his voice sound, why couldn't this illusion be applied to *things?*

Ironically, when it was decided to use sound effects on radio, it also was discovered that there was virtually no one in broadcasting with either the necessary experience or equipment. As a result, early radio again turned to the theater.

SOME EARLY THEATRICAL SOUND EFFECTS

Sound effects (SFX) have been providing illusionary sounds for audiences since the inception of the theater. As far back as the theater of Aeschylus, Euripedes, and Sophocles, the rumble of thunder was provided by "leaden balls, bounced on stretched leather, [simulating] the fabled wrath of Jove."

In 1708, John Dennis invented a new method of providing thunder for his play, *Appius and Virginia*. Although the play failed miserably, his thunder was a huge success. Dennis had used in place of the leaden ball and drum a sheet of copper suspended by wires. The effect of thunder was attained by holding the metal sheet by its edge and giving it an "appropriate shake" (see Figure 1–2).

The following series of illustrations and instructions as to how sounds were produced in the theater at the turn of the century are excerpted from the book, *Krows Equipment for Stage Production: A Manual for Scene Building*. Many of these sounds were duplicated and used by the early SFX artists in radio.

> Wind: *A very satisfactory wind can be produced by placing a heavy piece of canvas over a large wooden cylinder with an attached crank. By turning the crank in an appropriate manner, a very reasonable sounding wind can be produced. By turning the crank faster, these wind sounds can become of hurricane force.* [See Figure 1–3a.]

Wind sounds in the musical theater were most always done with a "bicycle siren provided by the trap drummer's outfit." (See Figure 1–3b.)

> Rain: *The sound of raindrops may be made with some dried peas or buckshot in a round, uncovered cheese-box. Another method is to roll peas or buckshot around in a bass drum that has had one drum head removed.*
>
> Surf or Waves: *Partly fill a long shallow box with loose shells or broken pieces of pottery. Cover the top with chicken wire. By placing the box on the top of a carpenter's horse and tilting it back and forth in a see saw motion, a very satisfactory surf and wave sound can be made.* [See Figure 1–4.]

As primitive as these sounds were, they served the theater very well. One reason for their success was the fact that the sounds were always accompanied by some visual special effect. If a scene called for a storm to

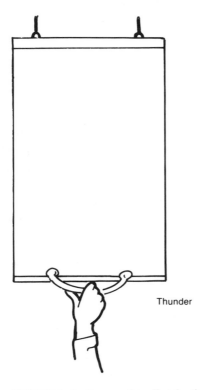

Thunder

FIGURE 1–2 Because this effect for thunder was so convenient and made such an impressive sound, soon other theaters began using Dennis's "thunder sheet." This so infuriated Dennis that his bitter complaint of "they're stealing my thunder" became part of our language.

be raging outside, the sounds of wind and thunder were accompanied by offstage fans blowing the window curtains while flashing lights indicated lightning.

Radio had none of these advantages and therefore the sounds had to be done in such a manner as to be self explanatory to the listening audience. As a result, the artists were always searching for new and better sounds. Figure 1–4 shows the number of artists it took to create a storm at sea using manual sound props from the theater.

Another valuable source of sounds and techniques was found in the musical theaters of vaudeville and burlesque: the house band's trap drummer. These talented musicians were called "trap" drummers because in addition to their musical function with the orchestra, these versatile gentlemen were required to have all sorts of small props, or trappings, on hand to satisfy any sound effect needs the various performers might have in their acts. Because of the popularity of this type of entertainment, the shows at

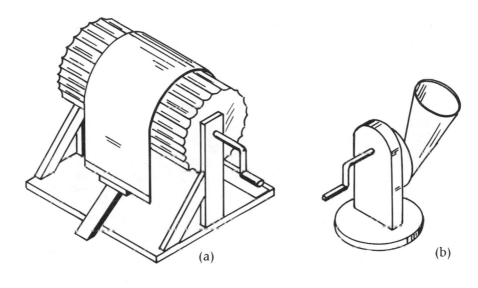

(a)   (b)

FIGURE 1–3

these theaters ran almost continuously, with little or no time to rehearse new acts. It therefore was not unusual for these drummers, in lieu of rehearsals, to receive such verbally descriptive instructions from the incoming new acts as: "Give me a cymbal crash and a bird twitter when I hit my wife in the rear end with my fiddle!" (See Figure 1–5.)

In order to accommodate these and other pictorially explicit requests, these drummers needed such sound effect trapping as wind whistles, metal and wood ratchets, slide whistles, train whistles, scratch boxes (for the sound of a steam engine chugging out of a station), fight gongs, blank pistols, and, of course, both a dry and water-filled, manually operated whistle for the sound of a "bird twitter."

Although very few of these multitalented musicians found radio alluring, they made important contributions in both equipment and technique (see Figures 1–6 and 1–7).

THE PIONEERS

While early radio owed many thanks to those trap drummers, the two people most responsible for developing SFX into an art form in radio and for training others were Ora Nichols, CBS, New York, and Lloyd Creekmore, KHJ, Hollywood.

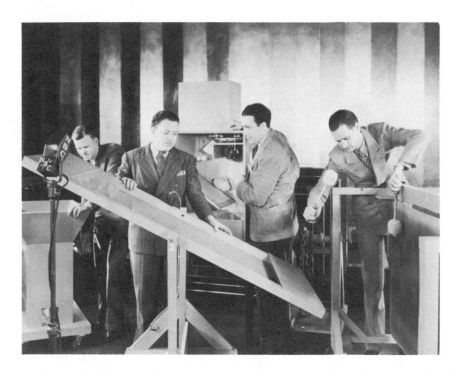

FIGURE 1–4 Pictured at the left is the wind machine. A heavy piece of canvas was placed loosely over a wooden shaft; when the artist turned the crank, it produced a very realistic wind sound. The long screen-bottomed wooden frame pictured in the center created the sounds of waves. By rolling BB pellets back and forth on the tilt-frame, the illusion of waves was created. On the far right is a thunder screen (see Figure 1–7). In the background is the rain machine. By turning a crank, a regulated amount of bird seed trickled from a hopper and fell onto a piece of stretched waxed paper. Needless to say, rainy days were not a favorite time for the sound effects artists in early radio. (Photograph courtesy of Virgil Reimer.)

It was a difficult job at best. Radio had very quickly outgrown many of the early drummers' traps because of their lack of realism. What had been appropriate for the large theaters was either too broad or unconvincing for the intimacy of radio.

During this transitional period, radio, out of a lack of technical equipment, began relying heavily on the actual effects for its sounds. If the script called for the sound of a coffee pot, the artist actually used a coffee pot. There were several reasons for this, the most important being that the artists still had not developed the art of successfully substituting one sound for another. It therefore followed that if the director saw the artist using a coffee pot for the sound of a coffee pot, how could he or she complain about the resultant sound? The problem was that in trying to satisfy all

FIGURE 1–5 Charles Forsyth stands amid his treasury of sound effects props in his Hollywood rental studio in 1933. In the background to the left is a thunder drum. In those days, when a radio show rented effects for a storm from Mr. Forsyth, the equipment literally had to be delivered to the studio in a truck. The majority of other rental sound effects in the studio once belonged to theatrical "trap" drummers.

these culinary script needs, the artist had to practically set up housekeeping in the studio!

The practice of using the actual effect to produce its own sound was both convenient and relatively successful in those early days. It was also only temporary. The demand for new sounds was increasing at such an alarming rate that the burgeoning shelves in the sound effects rooms began to fill and spill over (see Figures 1–8 and 1–9).

THE VISUAL DIRECTORS

The practice of supplying the actual sound for every needed effect was becoming increasingly impractical. And yet, what was to be done about the problem of satisfying the "visual directors"? A visual director was one

FIGURE 1–6 This is another example of a theatrical effect being adapted to radio. To the right is pictured a piece of wood used for the sound of horses on pavement or crossing a bridge. The gravel in the box was an improvement over carpet, but if the artist forgot to water it down before doing hoofbeats, the resultant dust cloud made it impossible to see the script or anything else in the studio—including the director. (Photograph courtesy of Ray Erlenborn.)

who only heard what she saw (see Figure 1–10). Perhaps it was their theatrical background, but they could never justify in their minds how the sound of coffee perking and lava bubbling could all come from the same pot. To them, a coffee pot was a coffee pot, and if you tried to use it for something else, it still sounded like coffee perking.

The effect for frying hamburgers was that of a wet cloth on a hot plate. However, one visual director insisted on using real burgers. Unfortunately, this was in front of a live audience. Also unfortunately, the show aired live around dinner time. The enticing aroma of the hamburgers set the audience off into envious murmurs that were almost louder than the actor's dialogue. From that time on, radio never had anything more exotic to barbecue than wet rags sizzled on a hot plate.

FIGURE 1–7 As sound effects artists became more familiar with the microphone, they began to replace many of the sound effects they had copied from the theater. Pictured here is radio's electronic answer to the thunder drum and thunder sheet. A contact microphone can be seen at the top and middle of the copper screening. The output of this microphone was amplified and fed to a speaker. By striking the copper screening with a mallet, artists were rewarded with not only thunder, but many other explosive sounds. This effect was invented by Stuart McQuade, NBC sound effects.

FIGURE 1–8 Monte Fraser of NBC, Hollywood, checks over some of the hundreds of manual effects that cram the shelves in the sound effects department. Although many of the effects pictured are identical to those found in Charles Forsyth's studio (Figure 1–5), there is one important difference: this picture was taken in 1968, almost thirty-five years later! As you can see, little has changed. Even today, many of these old manual effects are used in television and films.

SOUND EFFECTS BECOME RECORDED

The need for more sophisticated SFX that were beyond the capabilities of the manually produced sounds of the theater was becoming an increasing problem in the early 1930s. As a result, the artists began recording sounds on records. This was not, however, a total solution to the problem. Sound recorders in those days were approximately the same size and weight of a large washing machine. As a result, most of the sounds recorded were those that could be produced in the studio. Fortunately, large commercial companies such as Gennet, Standard, Speedy-Q, and Major also recognized radio's need for new and unusual sounds and adapted equipment that enabled them to go into the field to record realistic sounds. Because of the subsequent recording of actual sounds, radio drama was given another element of realism.

FIGURE 1–9 Lloyd Creekmore is shown in one of the studios at KHJ in Hollywood in 1938. Notice to the right and left of the studio the turntable consoles. Due to the tremendous popularity of radio sound effects, commercial companies were beginning to produce a variety of sounds on records that until that time could only be done manually or vocally. Due to the convenience and often superior sound quality of recorded sounds, this marked the end of such exotic effects as the thunder drum and that strange-looking box in the background with the antennas radiating from it—the sound effects artist's answer to a "wind of hurricane force."

In order to utilize these new sound effect records, phonograph record players were modified to accommodate the problems of the sound effects artist (Figure 1–11).

If, for instance, you needed the sound of wind behind a long scene, you simply played the wind sound with one pickup arm, and as the needle was coming to the end of the record you cross-faded the sound (slowly closed one pot while opening another) to the other pickup arm, which was placed at the beginning of the record. (See the center turntable in Figure 1–11.)

In addition to this "double-arming" capability, the speed of the motor could be made faster or slower than the normal 78 RPM speed; and this, perhaps, was the most important feature of these SFX consoles. By changing the speed of one sound, such as the "Mogambi Waterfalls," you could produce such varied sounds as gunshots, surf, traffic (add horns), a jet airplane, even the sound of the atom bomb exploding! (See Figures 1–12 and 1–13.)

RADIO'S INTENSE TEAMWORK

As the actors became familiar and more comfortable working with the SFX artist, they realized it was a method of reclaiming some of the theatrical techniques they had relinquished when they first entered radio.

FIGURE 1–10 This is the extent that some directors would go to for realism. Shown here, Walt Gustafson, ABC, New York, has an audience of two Pinkerton guards intently watching him make the sound of 100,000 dollars' worth of jewelry, with, naturally, 100,000 dollars' worth of jewelry. (Photograph courtesy of Anne Beach Gustafson.)

The famous stage actress, Eva LeGalliene, summed up her experiences by saying, "They were my stage actions, my slightest moves . . . they did everything I normally did on the stage except speak my lines."

The working relationship between the SFX artist and the actors became extremely close. If an actor was playing the part of an elderly man climbing a flight of stairs, the actor would supply the sighs and heavy breathing in sync with the SFX steps squeaking slowly up the stairs.

Just as the actors had found an ally in helping "flesh out" their characters, radio drama was given another dimension by SFX. In addition to the action and excitement SFX supplied, they helped end the confusion about the locale of a scene, the type of weather supposedly occurring, even the mood of the scene. But most of all, SFX helped relieve radio's incessant talking and allowed the audience to listen to the action and use its imagination. Radio, at last, had taken its first uncertain steps toward a unique identity . . . steps supplied by the SFX artist. (See Figure 1–14).

FIGURE 1–11 These consoles were further adapted to fit the requirements of sound effects. You will notice there are three turntables, and for each turntable there are two pickup arms. This gave the artist the ability to play one sound indefinitely. (Pictured is John McCloskey, CBS, New York.)

One of the most popular comedic sound effects in radio was the miserly Jack Bennys' frequent trips to visit his money in a subterranean depository referred to as "the vault." To get there, Benny was accompanied by numerous sounds on echo, including footsteps down stone stairs, clanging chains, metal doors squeaking open and closed, alarm bells, sirens, klaxon horns, and very often, sounds as sudden and unanticipated as a duck quacking frantically. Of course what added to the illusion of these sounds was Jack Benny's flawless timing. Interestingly, when this highly successful vault routine was tried on Benny's television series, it was a disappointing failure, the reason being that the reality of the vault set lacked the playful fun that radio audiences had imagined. In Figure 1–15, Benny's sound effects artist, Virgil Reimer, and the comedian indulge in a brief role reversal. What makes pictures like this so rare is that many radio executives felt that giving public recognition to the sound effects artist detracted from the mystique of radio.

FIGURE 1–12 Although commercial record companies were already supplying actual car crash sounds superior to any the artist could do manually, there simply was nothing funny to a studio audience about watching an artist playing a record. Comedy sound effects required movement and played off the reactions of the studio audience. Here, Virgil Reimer holds a metal "tub crash." Although small, it could be made to sound like Jack Benny's Maxwell car falling apart fender by fender, bolt by bolt . . . but always followed with the de rigueur of comedy . . . the tinkle of a small bell or clink of a teaspoon. (Photograph courtesy of Virgil Reimer.)

FIGURE 1–13 Paul Smith goes about the exacting task of creating the famous Fibber McGee closet crash. Each item selected not only had to sound funny, but it had to *look* funny for the benefit of the studio audience. Getting laughs from those few hundred people was extremely important not only to the comedians in the studio, but to the millions of listeners at home. The theory being, if the people at home heard people laughing in the studio, it must be funny. Even though this was in the live days before the invention of the laugh machines, many shows hired professional laughers called *claques*. Paul's assistant is Lisa Ingham. Photograph courtesy of the *San Francisco Examiner*.

SOME FAMOUS SOUND EFFECTS OPENINGS

Another function of SFX in those early days was to attract audiences. By cleverly blending exciting or unusual sounds into the opening of a show, listeners who were dialing through the stations in search of something interesting were often intrigued enough by the sounds to want to hear the rest of the program.

Although *Gangbusters* was by far the most famous, here are some other programs with some extremely popular and ear-catching openings.*

Superman

ANNOUNCER:
Faster than a speeding bullet!

---

*Reprinted by permission of Sterling Lord Literistic Ltd., New York. Copyright © 1966, 1972 by Bill Buxton and Bill Owen.

FIGURE 1–14 This picture was taken from the viewpoint of the director in the control room of Studio No. 1 at CBS, New York. As Frank Mellow cues up a record, Lloyd Morse prepares to strike the thunder drum. In the foreground, Jimmy Rinaldi turns a wagon wheel while Jack Amrhein works with marching feet. (Photograph courtesy of CBS.)

SOUND: GUNSHOT AND RICHOCHET
                    ANNOUNCER:
          More powerful than locomotive!

SOUND: LOCOMOTIVE EFFECT
                    ANNOUNCER:
          Able to leap tall buildings in a single bound!

SOUND: FLYING EFFECT, WIND UP FULL AND UNDER
                    MAN:
          Look! Up in the sky! It's a bird!
                    WOMAN:
          It's a plane!
                    MAN:
          It's Superman!!!!

FIGURE 1–15 In this brief role reversal, Jack Benny is being readied to do a gunshot by his sound effects artist, Virgil Reimer. Pictures such as this were never sent out to the radio listening public for fear that acknowledging the existence of sound effects on a show would disillusion their listening pleasure. It was for this same reason that the artists very seldom received any audio credits at the end of the program. (Photograph courtesy of Virgil Reimer.)

The opening for *Superman* is self-explanatory. In those days, even if you had never heard of *Superman,* you would be tempted to listen just a little longer to find out what all the excitement was about. That was the reason for hyping most openings with sound effects. Producers felt, as they do today on television, that if they could get an audience to listen to their program for thirty seconds, they would become interested enough to remain throughout the entire program.

Inner Sanctum
(Opening)

SOUND: DOOR SQUEAKS OPEN SLOWLY

> HOST:
> Good evening, friends. This is Raymond, your host, welcoming you into the Inner Sanctum. . . .

(Closing)

> HOST:
> Now it's time to close the door of the Inner Sanctum until next week. . . . Until then, goodnight . . . pleasant dreams . . . hmmmmmm?

SOUND: DOOR SQUEAKS SLOWLY CLOSED

Although the door squeaking open and closed on *Inner Sanctum* hardly sounds like anything that would attract listeners, the effect was done with such eerie slowness that it became one of the most recognizable sounds on radio. So much so that whenever an artist had to do the effect for a door squeaking open, the director would inevitably say, ''Give me something that sounds like *Inner Sanctum,* or ''I don't like it, people will think they're listening to *Inner Sanctum.*'' So unless you were doing *Inner Sanctum,* a door squeaking open wasn't a fun cue to do! (See Figure 1–16.)

The Green Hornet

> ANNOUNCER:
> The Green Hornet!

SOUND: HORNET BUZZ, UP FULL

> ANNOUNCER:
> He hunts the biggest of all game: public enemies who try to destroy our America!
> MUSIC:
> THEME (*Flight of the Bumblebee*), UP FULL AND UNDER

FIGURE 1–16   Ray Erlenborn, CBS, Hollywood. (Photograph courtesy of Ray Erlenborn.)

> ANNOUNCER:
> With his faithful valet, Kato, Britt Reid, daring
> young publisher, matches wits with the
> underworld, risking his life that criminals and
> racketeers, within the law, may feel its weight
> by the sting of . . . The Green Hornet!

SOUND: BLACK BEAUTY CAR PULLS OUT, UP FULL

> ANNOUNCER:
> Ride with Britt Reid in the thrilling adventure
> "Death Stalks the City." The Green Hornet
> strikes again!

SOUND: HORNET BUZZ, UP FULL

> ANNOUNCER:
> Stepping through a secret panel in the rear of
> the closet in his bedroom, Britt Reid and Kato

went along a narrow passageway built within
the wall of the apartment itself. This passage
led to an adjoining building which fronted on a
dark side street.

SOUND: FOOTSTEPS CROSS GARAGE FLOOR, FADE IN, AND CONTINUE
 THROUGH
>ANNOUNCER:
>Though supposedly abandoned, this building
>served as the hiding place for the sleek, super-
>powered Black Beauty, streamlined car of The
>Green Hornet.

SOUND: FOOTSTEPS TO CAR, DOOR OPENS, CAR DOOR SLAMS
>ANNOUNCER:
>Britt Reid presses a button....

SOUND: CAR STARTS
>ANNOUNCER:
>The great car roared into life....

SOUND: CAR IDLE UP, SOUND OF WALL SECTION OPENING
>ANNOUNCER:
>A section of the wall in front raised
>automatically, then closed, as the Black Beauty
>sped into the darkness.

SOUND: CAR PULL-OUT UP FULL, TWO GEAR CHANGES, CROSS FADE
 MUSIC

(See Figure 1–17.)

Gangbusters

>ANNOUNCER:
>Sloan's Liniment presents: Gangbusters! At
>war, marching against the underworld from
>coast to coast. Gangbusters, police, G-men, our
>government agents marching towards the
>underworld!

Punctuating the announcer's words were such sounds as glass breaking, police whistles, machine gunfire, sirens, and, of course, the marching
"feet." The combinations of these sounds were so exciting that even today
we have the expression, "coming on like Gangbusters!" (See Figure
1–18.)

And finally, the opening that audiences eagerly awaited from January
30, 1933, to September 3, 1954.

FIGURE 1–17 Some shows, despite how busy they were, only had a budget for one artist. Vic Rhubei, CBS, demonstrates how you do *"footsteps to car, door open, car door slam . . ."* followed by *"car starts, car idle up, sound of wall section opening. Car pull out, up full. Two gear changes . . . ."* Of course, that was just the opening.

FIGURE 1–18 This ingenious gadget was radio's "marching feet." The wooden pegs were suspended by wires or rawhide strings, and by rhythmically moving the frame up and down, a convincing "marching" sound was created. This effect was one of the sounds used for the opening of the famous radio show *Gangbusters*.

The Lone Ranger

MUSIC:
William *Tell Overture*, up full and under

SOUND: HOOFBEATS FADE IN

RANGER:
Hi-ho Silver!

SOUND: GUNSHOTS AND HOOFBEATS

ANNOUNCER:
A fiery horse with the speed of light, a cloud of
dust and a hearty hi-ho, Silver! The Lone
Ranger!

SOUND: HOOFBEATS, UP FULL

(See Figure 1–19.)

Buck Rogers in the 25th Century was a popular science fiction show long
before the term "science fiction" was part of the English language. First
heard over CBS (1931) and later over Mutual (1939), the problems this
show presented to the SFX artists were tremendous. It was their responsibility to come up with all the futuristic sounds that only existed in the
incredible imagination of the writers. Scripts abounded with such sounds
as sparkle static, repellent zap guns, dream water, and reducer rays. Naturally, all types of space vehicles were featured.

I have often thought how disappointing it would have been for a generation accustomed to listening to *Buck Rogers* on radio if when this country sent up its first space rocket it hadn't sounded like the old familiar
spaceships that Buck and Wilma had used to chase their nemesis Killer
Kane around the galaxy. Perhaps all systems were "go" for years. They
just didn't have the correct sound, in the *Buck Rogers* tradition, to accompany the blast-off.

The opening and closing of the show was a combination of a drum
roll on the sound effects thunder drums and the announcer's voice on
"echo," intoning, "Buck Rogerrrrs . . . in the 25th Century!" This echo
effect was created by the sound waves of the announcer's voice resonating
with the piano strings. (See Figure 1–20.)

THE THEATRE OF THE MIND

With the addition of sound effects, radio, at long last, had truly become
the "theater of the mind." The stage had its makeup, costumes, props, and
scenery; film had its camera, and now radio had perfected the art of SFX.

It was one thing for radio mystery shows to talk about being alone in
an old house on a stormy night with an insane killer on the loose, but it

FIGURE 1–19 "And as the Lone Ranger and his great horse, Silver, thundered across the western plains and plunged through deep mountain streams . . . ." Artists such as Monte Fraser and Virgil Reimer helped the listening audience to visualize the exciting action. The "western plains" were the two shallow gravel and dirt boxes in the foreground, while the "deep mountain stream" was the canvas-lined splash tank. Radio was indeed, the "Theater of the Mind." (Photograph courtesy of Virgil Reimer.)

was quite another to *hear* what it was like. The mantle clock striking the hour of midnight. A clap of thunder, rain beating against the window, a howling wind, shutters banging against the side of the house, and then, the unmistakable sound of steps coming slowly up the squeaking stairs. . . .

FIGURE 1–20 The unnatural, eerie, futuristic-sounding voice intoning: "Buck Rogggers, in the twenty-fifth cennntury . . ." was created by the announcer projecting his voice into this open piano, causing the strings to vibrate, thereby giving radio one of its first echo chambers. The combination of this unusual sounding voice and the rumblings of the sound effects thunder drum gave this program one of the most popular openings in radio.

Yes, at long last, radio had SFX, but more importantly, it had its own identity.

Radio Becomes Transcribed

In the days of live network drama, the three-hour difference between the Eastern and Pacific time zones created havoc for radio. Sponsors demanding that their program be heard at the most advantageous times in the large listening markets created the policy of doing the show live for the local time zone and then repeating it, again doing it live, three hours later for the other coastal time zone.

These three-hour delays in between shows were not only extremely tedious for all concerned, but they very often kept actors from accepting calls for other work. One actor was so much in demand that he hired an ambulance to get him through New York City's congested traffic.

Therefore, when the networks adopted a method of transcribing (recording) the programs on huge sixteen-inch acetates (records), everyone, especially the busy actors, was delighted (See Figure 1–21).

But as happy as everyone was with the convenience of these transcriptions, there were some farsighted members in the actor's union, American Federation of Radio Artists (AFRA), who also saw the dangers. They insisted that the actors be paid for these transcribed shows, even though they were not present when the shows aired. Quite a revolutionary concept in those days. A concept, I might add, that was the groundwork for today's lucrative residual market.

The End of Radio Drama

As the practice of transcribing shows became increasingly popular, it marked the beginning of the end of live dramatic radio, just as later, video tape would sound the death knell for live television.

Although these transcribed shows had many advantages over live productions, they also had disadvantages. As previously stated, transcriptions, unlike audio tape, could not be edited for mistakes. Once a show was started, it had to be completed in one take.

On one memorable Christmas show, the SFX artist had to do the manual hoofbeats for the donkey carrying Mary to Bethlehem. These sounds began on page 1 and lasted for 28 of the 34 pages of the story. Unfortunately, the only way to make the hoofbeats was manually with coconut shells and a tray of sand: a very tedious and demanding effect, especially if it had to be done for long periods of time (see Figure 1–6).

FIGURE 1–21 The cast of *Gangbusters* listens to a transcription of program corrections and suggestions made by the producer, Philip Lord. What made this arrangement so unusual was that the cast of the popular crime show was in New York, while Lord was on his island hideout in Maine. After each weekly rehearsal, the director would send a transcribed recording to Maine, and in return, Lord would send back a transcribed recording of his critiques in regard to everything from script cuts to changes among cast members. In this manner, Lord kept four highly successful radio shows going weekly. On the far left are the sound effects artists Harry Nelson and Bob Prescott.

During all of the rehearsals, the director insisted on hearing the SFX of the hoofbeats, so by the time it came to record the show, the artists' hands were already tense and tired. To make matters worse, one of the actors with a leading role was a celebrated film star, but totally unfamiliar with the rigors of doing a radio show in one take. As a result, when the recording began, he got only as far as page 2 before making a mistake, thus necessitating scrapping the recording and starting all over. On subsequent takes, the story got as far as pages 10, 14, and 18 before a mistake occurred, requiring starting over again.

One take went as far as page 28 before a halt had to be called! By the time a record was finally cut that met the approval of the director, the artist's hands were stiff and numb from doing thousands upon thousands of hoofbeats.

In addition to the physical fatigue, it became very nerve-wracking for everyone concerned to work under these stressful conditions, especially for the actor who had a difficult and emotionally draining role. As the script neared completion, the actor would cross his fingers and pray that nothing would go wrong so he wouldn't have to repeat that demanding scene. When mistakes did occur requiring a repetition of the scene, tempers sometimes became very short.

Interestingly, many of the stoppages were due to insignificant mistakes—mistakes that, had they been done on live radio, probably would have gone unnoticed by everyone, including the home audience.

Transcribed shows gave birth to another increasingly annoying habit on the part of directors. Shows that were once considered superior by live standards were now being subjected to a new standard of excellence. Thus the expression, "That show was just excellent everybody, but I think we can do it a teensy trifle better." With the utterance of that single sentence, an era came to an end. Today it is called The Golden Age Of Radio, but back then, it was simply called live radio.

A HISTORY OF TELEVISION SOUND EFFECTS

The history of television goes all the way back to 1884, when Paul Nikow developed a mechanical scanning disk. However, it was not until 1948 that public interest in this revolutionary medium began. The person credited with television's sudden popularity was a brash young nightclub comedian by the name of Milton Berle. The show was called *The Texaco Star Theater*, and it was an instant hit. Even people who didn't have television sets would huddle outside an appliance store window and watch the program on a set provided by the enterprising owner. While Milton Berle was referred to as "Mr. Tuesday Night," over at CBS on Sunday nights, Ed Sullivan was starring in his own variety show called *Toast of the Town*. As the public's interest in television suddenly caught fire, networks began the frantic search for new types of programming and talent. It was at this time that the "Golden Age of Television" was born.

A Familiar Story

Just as radio had turned to the legitimate theater for studio space, so did television. Only now, because radio still occupied many of the theaters, television began leasing many of the large movie palaces that ironically were available because of the sudden competition from television.

Technical crews ignored the baroque splendor of these venerable palaces that had been specifically designed to give a feeling of opulence to a depression-weary public. Power lines, camera cables, microphone cables, lighting lines, air conditioning cables, and just plain lines and cables criss-crossed each other in a cobweb of confusion. Control rooms were installed, klieg lights hung, microphones put in place, and cameras rolled in. The anachronism of the gilded cherubs that adorned the proscenium arch and the electronic cameras that now peopled the stage was startling. Fred Allen, perhaps radio's most gifted comedian, summed it up succinctly: "Television is a triumph of equipment over people. . . ."

The Early Studios

The tremendous amount of lighting needed for those early cameras made the studios suffocatingly hot. No one was spared from the glaring heat, but if anyone got the brunt of it, it was the actors on camera. Makeup people and hairdressers hovered behind the cameras, waiting for those brief off-camera moments to make hurried repairs on hair and to mop heavily perspiring faces.

Not even actors with poor eyesight were spared. No matter how essential eye glasses were to an actor's vision, they couldn't be worn while on camera. If the glasses were essential to a star's public image, the frames could be worn, but without the lenses. Anything that caused reflections that would hurt the camera was forbidden, including windows, mirrors, and even the diamonds in the necklace around the star's neck. Everything was considered expendable if it interfered with the camera getting pictures.

Sound Effects Are Ignored

In those early days of television, sound effects were the very least of the concerns that plagued the network engineers and members of the production companies. Good pictures and hearing the actors say their lines received top priority, and everything else was secondary, certainly sound effects. Although networks were spending huge sums of money on equipment that affected the picture and acoustical quality of these newly acquired "studios," SFX were all but ignored. This was by no means intentional on the part of the networks; it's just that SFX were the very least of the things that were broken, so why fix them?

As a result, the equipment used by the artists was the same equipment used in radio. Even the effects were the same. Once everyone understood that the actors could do their own sounds when they were on camera and SFX artists did all off-camera sounds, everyone went back to the real problem of getting good pictures and hearing the actors' lines—except the stage hands union, International Alliance of Theatrical Stage Employees

(IATSE). Inasmuch as IATSE always had jurisdiction over SFX done on the stage, and these converted theaters were still technically considered "stages," they saw no reason why their electric department shouldn't do the effects.

It certainly wasn't because their electricians were not capable of the job, because at the time of these union disputes, about the only effects television was using were of the doorbell/phone bell variety. However, it was the presence of the camera and microphone that gave the broadcasting unions (the National Association of Broadcast Engineers and the International Brotherhood of Electrial Workers [NABET and IBEW]) jurisdiction over what was now referred to as a "television studio at a remote location."

One producer's response to the arbitration furor that was being paid to SFX was succinct, if not too knowledgeable: "What's the big deal over sound effects . . . we're doing television . . . not radio!"

Of course, the people who really understood SFX were still doing radio and were convinced that television was just a momentary diversion on the part of the public. Directors from the theater, unaccustomed to anything more complicated than an offstage door knock or phone ring, assumed that was all that was needed in television as well. The few experienced people from film who entered television perhaps had the most difficult time. Accustomed to the world where mistakes and SFX were both taken care of in post production, they, too, largely ignored the potential of SFX in the live studio. Even the young people who entered television without prior experience in the theater or films found the job of getting a program on and off the air in the allotted time and still getting good pictures and understandable dialogue so challenging that SFX were considered a luxury they could not afford.

As a result of television's ignorance, the areas assigned to SFX were more of an afterthought than one of acoustical design. Sound effects for such popular shows as *Danger, Suspense, The Web, Omnibus,* and *Studio One* were done high above the studio floor on a narrow lighting bridge at CBS Studio No. 1 at the famous Grand Central Station, which also served as one of the busiest railroad terminals in New York City. (See Figures 1–22 and 1–23.)

Simply because television did not understand SFX does not mean the effects were not appreciated. And just because the effects weren't written in the script didn't mean the artists didn't provide the necessary effects to add extra production values to the show. Perhaps that was the reason for sound effects being ignored in those early years. While everyone was so preoccupied with the technical and production problems associated with getting good pictures and audio, SFX artists were, in essence, still doing what they had always done for radio, providing imaginative sounds.

FIGURE 1–22 One of the many adjustments that radio sound effects artists had to make when they entered television involved the loss of the camaraderie that existed between the artists and the performers in radio. Here, Keene Crockett jokes with Jerry Colonna and Bob Hope. (Photograph courtesy of Keene Crockett.)

A HISTORY OF FILM SOUND EFFECTS

The silent film era is acknowledged to have its commercial start in this country on April 23, 1896. At that time, at Koster & Bial's Music Hall in New York's Herald Square, there was a series of Edison short films shown on the screen at Armat's Vitascope projector.

The success of the "moving picture" form of entertainment was immediate. Although subsequent films were little more than a novel and convenient method of presenting variety acts, there was a receptive audience wherever they were shown. And in those days, this included the tents of traveling carnivals, penny arcades, or even empty stores with borrowed chairs for the customers to sit on.

In 1903, Edwin S. Porter, a former Edison cameraman, broke with popular tradition and produced a film that was unique in that it had a story the audiences could follow. The film, *The Great Train Robbery*, was greeted with overwhelming success and generated a tremendous interest in this new and exciting art form.

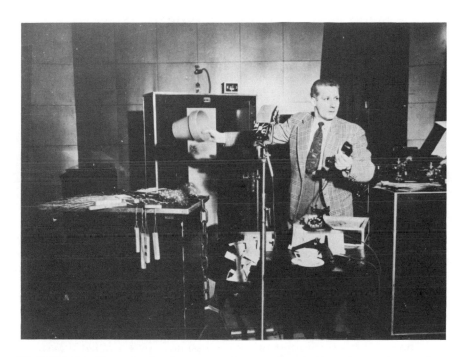

FIGURE 1–23 Keene Crockett in a more serious mood on the Hope show, as he gets down to the business of getting laughs. (Photograph courtesy of Keene Crockett.)

The Nickelodeons

One such enthusiast was John P. Harris. In 1905, he used his vaudeville theater in Pittsburgh, Pennsylvania, to show a series of motion pictures. Charging an admission of five cents—a nickel—he advertised his program as the "Nickelodeon."

The idea for this type of entertainment soon captured the imagination of entrepreneurs throughout the country. For a modest investment of approximately $150 dollars and a place for people to sit down out of the elements, anyone could become a film exhibitor. Although the studios originally sold the films outright to the exhibitors, this practice was abandoned for a rental fee charged by film exchanges, or libraries. As might be expected, unscrupulous exhibitors soon began "duping" (copying) films, saving themselves the exchange fees. Despite these practices, the film studios were showing enormous profits.

Perhaps it was because of these profits that Hollywood film makers were so reluctant to change their methods of operation, yet there were some warning signs for Hollywood's beloved "silents."

Some Warning Sounds Are Heard

Despite the fact that in 1925 only ten percent of the homes in America had radios, its sound was already beginning to exert a tremendous influence on its listeners. No longer did the American public have to rely on their newspapers for the latest news; now they could listen to it on their radios. In addition, there were informational programs on everything from how to raise your baby to making the best of a bad marriage. And perhaps best of all, there was music! Everything from *The Cliquot Club Eskimos,* to the great Enrico Caruso singing *Vesti La Giubba,* live from the Metropolitan Opera House. All in the comfort of the listener's home! There were warning signs all right, but with the exception of one studio executive, Sam Warner, everyone chose to play a waiting game with America's fascination with sound.

In 1925, Sam Warner went to New York to witness a sound demonstration given by Bell Laboratories that convinced him that sound was the future of the film business. The film he saw showed a small orchestra playing a song that was in such perfect synchronization with the music being heard over the loud speaker that the incredulous Sam Warner ran behind the screen to see if there were not live musicians playing.

Upon his return to Hollywood, Warner explained to his brothers what he had witnessed. After much discussion, it was decided they would sign an agreement with Western Electric, a subsidiary of Bell Laboratories, for their sound system.

This system, called Vitaphone, involved two synchronous motors. One motor moved the film through the camera while the other motor turned the turntable that played the sound disc in sync with the picture.

On August 6, 1926, the Warner brothers released the full-length movie *Don Juan.* Although this film was shot silent, Warners added a musical score recorded on disc by the New York Philharmonic Orchestra. This ambitious use of music synced to a film sparked a sudden interest in the commercial possibilities of sound in the film industry.

THE IMPACT OF SOUND

Much has been written about the dread the early silent screen stars had for sound. The fact that some of the stars did have unacceptable speaking voices is obviously true. However, what most of the stars had to overcome was far more difficult, because it was of a psychological nature.

We spoke of how the physical appearance of the radio actors was of little importance if they were able to conjure up the proper image with nothing more than the sound of their voices. Just the opposite was true of the silent film stars. Here they were selected from all over the world, regardless of their vocal quality, just as long as they looked the part they portrayed on the screen.

There existed among fans a tremendously mystical, almost religious adoration for the gods and goddesses of the silent screen. In their fantasies, they knew what their favorite screen stars *should* sound like. Therefore, it wasn't so much that the voices of those early silent stars were unacceptable for sound as it was that they were inconsistent with the expectations of fans and critics. It was this feeling of betrayal more than the actual voices, more than anything else, that sent the careers of so many silent screen stars tumbling from the heavens over Hollywood.

Contrary to what many believe, the silent films were not a crude form of entertainment that suffered because of lack of sound. They were a highly developed art form, drawing on the pantomime talents from artists throughout the world. Sound—especially dialogue—was considered totally unnecessary. Whatever music a film needed was supplied by the musicians in the theater. And if a situation needed sound effects, there was always the reliable, multitalented trap drummer or theater organist.

Most people in Hollywood who worked in films supported these anti-sound views. Charlie Chaplin, the great silent screen pantomimist, perhaps fearing what the addition of dialogue would do to his beloved *Little Tramp* character, had this to say about sound.

> *They are spoiling the oldest art in the world—the art of pantomime. They are ruining the great beauty of silence. They are defeating the meaning of the screen.*

A Worldwide Problem

Although much has been written about the impact that sound had on the Hollywood community, they were not the only ones who mourned the loss of the silents. In Japan, for instance, instead of showing written subtitles on film to forward the story or explain a scene, the Japanese had a man seated on the stage, off to one side of the screen, who vocally described the film. This highly skilled gentleman was called a *benshi*. Although his job was primarily that of an actor, he was also required to supply SFX vocally. In one scene he might provide the voice of the outraged feudal lord, weep hysterically for the wrongly accused wife, and somehow manage to do the barking for the pet Pekingese. Other vocal effects required of the benshi were as varied as a lion roaring, a monkey chattering, and cannons firing.

So popular were these vocal interpreters of the silent films that audiences often came to the theater to watch and listen to the benshi, regardless of what was playing on the screen. With the introduction of sound, the benshis remained in their customary chairs and tried valiantly to compete with the actors' voices and the SFX coming from the screen. But even the beloved benshis were no match for the magic of Hollywood's "talking picture."

Some Problems With Sound

In addition to the well-publicized problems Hollywood was having finding actors with suitable voices for sound, the studios were facing other equally serious challenges—learning how to deal with the sound microphone itself.

Unlike the good old days when one visitor to a silent picture set characterized it as, "probably the noisiest place in Hollywood," this new "talkie" demanded complete and utter silence . . . and this meant everyone except the person who was talking! Directors accustomed to coaching their actors through a scene for a silent picture were now forced to watch in grim silence. Even the cameras had to be enclosed in a box to prevent the motor noises from being picked up by the microphone. And microphones were everywhere! Hanging from the ceiling, concealed in a bouquet of flowers, or even hidden in a hat on a coat tree—just as long as they were near the actors and hidden from the camera.

Sound even made the camera operator suffer. Enclosed in a stifling box, the operator could shoot from only a limited number of angles. With all of these problems, it is no wonder so little attention was paid to SFX. If it wasn't heard over one of the actors' dialogue microphones, chances are it wouldn't be heard at all—or at best, poorly.

In the movie *On Trial*, reviewers took note to criticize the sound of a gunshot: ". . . that instead of adding to the suspense of the scene, the sound of the gunshot actually broke the mood by sounding about as loud and threatening as a peanut being snapped."

As the public's interest in sound increased, their level of tolerance for careless or inept sound techniques decreased; and this included SFX. Seeing a rooster on the screen and hearing a rasping noise that sounded only remotely like a rooster crowing was becoming increasingly unacceptable (as were inferior "gunshots").

Sound Leaves the Studio

While Warner Brothers was busy promoting its disc recording Vitaphone sound system, Fox Studios decided to adopt a new optical film method for their entrance into sound. William Fox, the studio head, named it Movietone.

Because of the greater mobility that the optical film method provided over the more cumbersome and fragile disc-cutting equipment used by Vitaphone, the Movietone cameras successfully took sound out of the studio.

In May of 1927, theater audiences were astounded to see and *hear* a filmed account of Lindbergh's takeoff for Paris, actually hearing the engine sounds of the *Spirit of St. Louis.*

Although Movietone's early successes were in newsreels, in 1928, Movietone filmed its first all-talking feature, *In Old Arizona,* outdoors. For the first time, SFX were applauded for their ability to "inject a sense of stark realism to the film." Singled out were the sounds of a ticking clock to heighten the tension of an impending gunfight and the sizzling sound of bacon frying.

In 1929, King Vidor made an arrangement to rent the Movietone camera and sound system for his movie, *Hallelujah.* Anxious to begin shooting, King Vidor learned there would be a delay in the arrival of the Movietone equipment. Rather than have his large cast and crew waste valuable time in the uncomfortable surroundings of an Arkansas swamp, King Vidor decided to go ahead and begin shooting without sound. As a result, much of the film's sound was done in post production.

The treatment that Vidor gave SFX is of special interest. It is probably the first example of SFX being used in place of music to arouse emotions. In a scene where the hero is being chased through a swamp, all the sounds are exaggerated to emphasize the nightmarish horror being experienced by the hunted man. Branches tear loudly and cruelly at his clothes; the thick mud sucks at his feet, making it difficult to run; birds shriek, and the hounds bark and howl viciously.

This film was a landmark of achievement for the art of SFX. King Vidor, in graphically demonstrating that the source of a sound is far less important than its quality, led the way for other directors to stop slavishly trying to record every little sound simply because it was natural. Even today, that is excellent advice.

SUMMARY

1. Although the first commercial radio broadcast was in 1922, it was not until the early 1930s that the Golden Age of Radio began.

2. Prior to the 1930s, the few SFX that were used in radio were done by the actors and other production personnel. Due to their lack of skills, writers had to include such descriptive dialogue as, "I think I hear a horse coming!"

3. One of the biggest fears of early radio was silence, or "dead air." With the development of SFX as an art form, radio had a welcome alternative to actors constantly talking.

4. The two people most responsible for the refinement of SFX as an art form were Ora Nichols, CBS, New York, and Lloyd Creekmore, KHJ, Hollywood.

5. Although the early SFX artists came from all walks of life, they depended heavily on the equipment and techniques developed by the theater's trap drummers.

6. As the demand for SFX in radio increased, new and more realistic SFX were needed. Inasmuch as many of the sound devices were built by the artists themselves, a busy SFX show might literally require a truckload of props.

7. Many of these SFX props, despite their unusual appearances, produced very realistic sounds. However, because many directors only heard what they saw, many props were turned down simply because of their appearance.

8. Because of the tremendous demand for new and better sounds, commercial recording companies began supplying artists with some desperately needed records of sounds actually recorded at the source.

9. Despite the availability of realistic sounds on records, comedy show directors often demanded that the artist use old-fashioned or outlandish props for their sounds in order to elicit laughs from the studio audience.

10. Radio required such intense teamwork from musicians, actors, and SFX artists that many listeners at home were totally unaware there was such a thing as "sound effects."

11. One of the popular usages of SFX was for the opening of action stories. Some of these "teasers" became so popular that they actually overshadowed the program itself.

12. In the late 1940s, many producers decided it was more economical and convenient to put their programs on transcription records, thus ending the exciting and creative era called "live radio."

13. So great were the technical problems involved with getting good pictures and audio in television that the need for SFX was greatly ignored.

14. Sound effects for early television were created using the same techniques and equipment used by radio artists. The only difference was the amount of effects involved.

15. Although early films are called "silent," they used both music and SFX provided by the musicians in the theaters.

16. Much of early filmmakers' knowledge about sound was provided by engineers with experience in radio.

17. Early SFX and music for films were provided by discs played in the projection booth.

18. In 1929, when King Vidor produced his film *Hallelujah*, the SFX were added to the film in postproduction. So successful were these sounds in creating a desired mood that for the first time, SFX were used in place of music. This film marked the end of SFX being used in film simply for their novelty value.

CHAPTER 2

SOME BASICS OF SOUND

A HISTORY OF SOUND

As far back as 500 B.C. the ancient Greeks were experimenting with sound. Pythagoras of Samos discovered when plucking the strings of a lyre that the shorter strings produced a higher sound than that produced by the longer strings. In addition, when a string was plucked in unison with another string twice its length, it produced a very pleasant sound that differed by an octave. Furthermore, by conforming to other simple ratios such as two to three or three to four, other combinations of strings produced equally pleasant (harmonious) sounds. It was this application of whole numbers to the lyre's strings that first caused the Greeks to include music with other mathematical subjects such as geometry, astronomy, and arithmetic.

Archytas of Tarentum, in about 400 B.C., theorized that sound was produced by striking objects together. He also stated that objects struck together sharply produced a sound higher than those objects struck with a slower motion. Aristotle, in about 350 B.C., stated that air contributed to sound production. He theorized that the air next to the struck bodies moved, striking adjoining air, thus propogating sound. Further, without a medium of propagation such as air or water, man would not be able to hear sound.

Much later, this theory was proven with a simple experiment. An electric bell suspended on a band of rubber and lowered into a large glass jar containing air can be activated and readily heard ringing. However, when the air is withdrawn from the jar, the ringing is no longer heard, but the

40

clapper on the bell can be seen moving as vigorously as before. By lowering the bell so that it rests on the bottom of the airless glass jar, the sound can once more be heard, proving that the glass and the wood of the table act as conductors for the sound. It was therefore concluded that gases, liquids, and solids can conduct sound, while a vacuum cannot.

THE ELECTRON

The Greek word for *amber* is *electron*. It was discovered that when amber was rubbed with a woolen cloth it became electrostatically charged and attracted other small objects in much the same manner as a magnet. This phenomenon occurred due to electrons being removed from the cloth and added to the amber. The important step was the dislodgement of electrons by the rubbing process, which created the necessary heat. Perhaps a more familiar example of static electricity occurs when you walk on a thick carpet and touch a doorknob. The "shock" that you receive is a flow of electrons between you and the doorknob. The flow of electrons through a conductor is electricity; the flow in turn produces an electromagnetic field.

Electrons are the negatively charged particles in the structure of atoms, which in turn are the smallest material units that make up matter.

At the nucleus (center) of an atom are two other particles that remain stationary. One is a positively charged particle (proton); the other is a particle with no electrical charge (neutron).

Electrons orbit the nucleus of an atom in the same manner that planets revolve around the sun. In addition, like the planets, the electrons also rotate on their axis.

When an atom lacks orbital electrons to form a balance with its protons, it will attempt to secure this balance by acquiring other electrons.

Particles that contain a deficiency of electrons are said to be positive (+), and particles with an excess of electrons are said to be negative (−).

Before you attempt to visualize how small an electron is, it should be noted that it takes 31,000,000,000,000,000,000,000,000,000,000,000 (31 octillion) of them to weigh one ounce. Because of this, the electron's characteristics are determined by its electromagnetic and electrostatic effects rather than by observation.

VISUALIZING SOUND

All liquids, solids, and gases are made up of atoms. Atoms that are arranged in groups form elements called molecules. It is the vibration of air molecules that causes the phenomenon called sound.

To make this easier to visualize, think of a small pool of water. When an object is dropped into the water, it disturbs the water's molecules and we see the familiar splash. The size of the splash and the resultant waves are dependent upon the size of the object and the manner in which it enters the water. Is it a pebble or a large rock? Is it dropped gently or thrown with force into the water? All of these factors will determine the size of the splash and the magnitude of the waves. Waves are measured from their peak to their trough (depth) and continue out in a circular fashion until the wave's energy is finally dissipated.

Transverse Waves

Although the analogy of water to a sound wave is helpful in visualizing the effect of a wave of sound vibrating from a source, it isn't an entirely accurate one.

When an object disturbs the surface of a pool of water, waves fan out in a circular pattern. However, these waves do not move away from the disturbance; rather, they move in an up and down fashion. The waves closest to where the rock entered the water will have the greatest amplitude, while the waves in the subsequent outer circles will have less height. Because these waves move in an up and down movement, they are called *transverse waves*.

This wave motion can be seen if you watch a small boat in a sheltered harbor. The movement is that of bobbing up and down, and unless there is wind or a current, the boat will remain in that relatively same position.

Longitudinal Waves

Sound waves, although similar in appearance, move in a different fashion. When a vibrating object (sound source) disturbs the air molecules, it forces them together and compresses the air molecules in the immediate vicinity. These compressed molecules move away as a result of the greater pressure exerted by the neighboring air molecules. In turn, these neighboring molecules are "bumped" together, forming another compressed grouping of molecules; on and on this phenomenon continues until the sound waves run out of energy. These sound waves, because they move away from the vibrating object, are called *longitudinal waves*.

THE SOUND WAVE

When a sound is created, the atmospheric molecules react in a similar fashion as the molecules of water when its surface was disturbed by an object. That is, the molecules in the atmosphere are compressed to form a wave of sound. Just as water has peaks and troughs, so does a sound wave; however, in sound, the troughs are called *rarefactions*. When a sound disturbance or vibration has caused the compression of molecules from peak to rarefaction, it is said to have completed one *cycle*.

The sound wave is composed of a series of peaks and troughs moving in alternate directions. However, once they have caused the adjacent molecules to be in a compressed condition, they return to their original position due to the *elasticity* of the air.

In Figure 2–1, as the bell begins to ring it disturbs the molecules of air and forces them to alternately bunch up (compress) and spread apart (rarefaction). As the compressed air moves to fill the lower pressure area (rarefaction), a sound wave is created. The farther these waves travel from the bell (sound source), the fainter the sound will be.

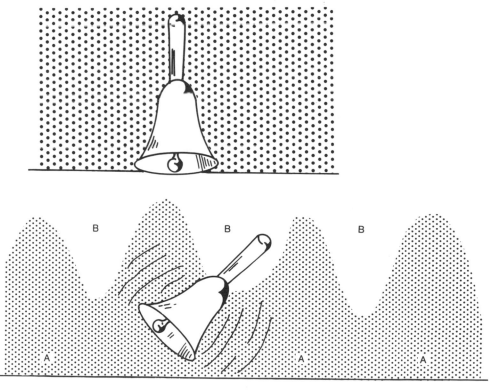

FIGURE 2–1 (Top) Bell at rest surrounded by molecules of air. (Bottom) The ringing bell causes the molecules of air to compress (A) and to draw apart (B) (rarefaction).

If you were to suspend a weight on the end of a resilient spring, pull the weight, then release it, the weight would move up and down, alternating above and below its original resting place, or position of equilibrium. How long this weight stayed in movement would depend a great deal on the ability of the metal spring to "want" to snap back to its point of equilibrium. The ability of the spring to accommodate these various forces is the elasticity of the spring. Good conductors of sound have this same ability.

The Sine Wave

Figure 2–2 illustrates a pen attached to one side of a tuning fork. When the tuning fork is made to vibrate, a tape of paper moves from right to left, and as it does, the pen imprints a wavy line called a sine wave on the paper. This sine wave is the fundamental wave of a sound.

Frequency

When a sound disturbance occurs, it sets up a series of sound waves called cycles. The number of cycles that occur in one second is said to be that sound's *frequency*. If a sound causes vibrations to complete 500 cycles per second, it is said to have a frequency of 500 cycles per second, or as it is written today, 500 hertz (Hz). The term *hertz* was adopted to honor the nineteenth-century physicist Heinrich Hertz.

Human beings are capable of hearing sounds within the range of 20 hertz (Hz) to 20,000 Hz. These numbers, as we shall learn, are somewhat idealized and are subject to a number of conditions such as age, sex, and length of exposure to loud sounds.

Wavelength

A wavelength is the measurement of one complete cycle of sound. If an object causes the air to oscillate (vibrate) at the rate of 250 cycles per second, the frequency of those vibrations is said to be 250 Hz. If these vibrations were a pure tone, meaning a tone without harmonics or over-

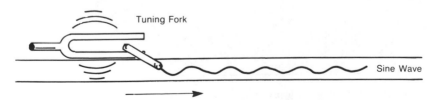

FIGURE 2–2 A sine wave generated by a pen attached to a tuning fork.

tones, this single frequency of 250 Hz could be measured on a graph as a *sine wave*. An object capable of producing a single frequency, or sine wave, is a tuning fork.

If we were capable of taking this tuning fork and measuring its vibrations on a length of graph paper, in order to find the wavelength of these vibrations, the paper would have to travel at a rate of 1,100 feet per second, or the speed of sound.

There is a simpler way of calculating the vibrations mathematically. The wavelength of a sound is determined by dividing the velocity (speed) of sound by the frequency of the sound. If the frequency is 100 Hz, we divide that number into the velocity of sound, 1,100 feet per second; this gives us an answer of 11 feet. Therefore, the wavelength of a cycle of sound at 100 Hz is 11 feet.

Using our water analogy, suppose that we had the ability to suddenly freeze the water when its surface was disturbed. We would then be able to see and even count all the vibrations that the disturbance caused.

Figure 2–3 represents a series of sound vibrations. Distances from points A to B, B to C, C to D, and D to E are all measurements of wavelengths—one complete cycle of a wave. As we look at all these frozen waves, what do they tell us? Absolutely nothing. In order for frequency and wavelength data to be meaningful, we must have a time factor. Regarding Figure 2–3, if it takes 1 second for those cycles to occur, that drawing represents a frequency of 4 Hz. Once again, using our mathematics, the velocity of sound (1,100 feet per second) divided by a frequency of 4 Hz gives us a wavelength of 275 feet for each cycle illustrated in Figure 2–1b.

Although it is impossible to freeze sound waves, we can do the next best thing: record them on tape. If we record a 100-cycle tone at a certain speed and play it back at the same speed, we know that each cycle of sound is passing the playback head at a rate of 100 times per second. To put it another way, each cycle takes 1/100 of a second to pass the playback head. We also know that the wavelength of each cycle is 11 feet.

But now suppose we take this same tape of the recorded 100-cycle tone and play it back exactly half as fast. Instead of each cycle going past

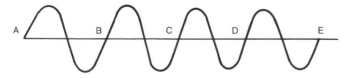

FIGURE 2–3 A series of sound vibrations. Points A to B, B to C, C to D, and D to E represent wavelengths, or complete cycles of a wave.

the playback head at a rate of 100 times per second, it is only going past at a rate of 50 times per second. What does that do to our wavelength? Nothing—it remains unchanged—because we are not changing the frequency of the recorded tone, only the speed in which it is being played back. Therefore, although the wavelength does not change, the playback head of the tape machine "reads," instead of 100 cycles passing its head in one second, only 50 cycles in one second, and the frequency now heard is exactly one octave lower.

This technique of playing sounds back at speeds different from their recorded speed is another effective method of creating new sounds.

Velocity

The velocity of sound in air is approximately 1,100 feet per second or 740 miles per hour. The vibrations of sound will continue until the sound source ceases, and then the waves will be dissipated or diffused by the atmosphere.

The speed of sound is slower than the speed of light. This is readily illustrated on a stormy night when you first see a flash of lightning and then hear the sound of the thunder. Other examples of this phenomenon can be observed with gunshots. First you see the flame and smoke, then you hear the shot sound. The farther away the gun is, the greater the separation between what you see and the sound you hear.

Another aspect of sound that affects its velocity is temperature. As the temperature rises, the air becomes more elastic and, therefore, a better conductor. That is why on a hot summer night it isn't simply because you have your windows open that you can hear your neighbors arguing or playing their music seemingly so loudly, it's because sound travels so much better in warm air. Conversely, as the air cools, the velocity of sound decreases. The ratio is as follows: for every change of 1°F, the velocity of sound changes 1.1 feet per second.

ECHO AND REVERBERATION

A sound wave is not unlike a ray of light. It can be intensified, weakened, enlarged, shrunk, refracted, and more.

When a vibration occurs in a room, the person present hears:

1. the original or direct sound,

2. the early reflections of that sound called echo, and

3. the later reflections of that sound referred to as reverberation

All of these variables are dependent upon the size of the room and the position of the listener.

Direct sound refers to the sound that the listener hears from the source without any delay. It is this sound that helps the listener identify a sound's direction.

Echo is caused by the direct sound being reflected off a nearby surface. The less absorbent the material, the more distinct the echo.

Reverberation of a sound is the later reflection of a direct sound. As these reflections are repeated with a very short time interval, they create reverberations; therefore, we can define reverberation as a series of echoes so closely spaced that they overlap. Whereas direct sound gives us information as to the location of a sound, reverberation gives us such spatial information as to the approximate size of a room. As the energy of the reverberation weakens, the sound will die away, or decay.

The decay time for various sounds varies widely. However, it is an integral part of the sound. If a sound is not allowed to decay properly, it will have an unnatural clipped sound as if the sound has suddenly been cut off.

Reflected Sounds

Although the study of reflected sound is relatively new, many species of mammals communicate using it and depend on reflective sound to exist. Bats, for instance, emit ultrasonic sounds of up to 150,000 Hz. These ultrasonic sounds reflect off flying insects, thus enabling the bats to find food, and off obstacles such as trees, thus helping them to navigate. Porpoises emit the same type of reflective sounds in the water.

To a lesser degree, humans depend on reflective sounds. For example, by tapping a cane, unsighted people can determine whether they are between two buildings or in an open space. This is accomplished by listening to the reverberation of the taps. The more echo or reflection the sound has, the more contained the area is.

Acoustics

The science of absorption and reflection of sound is called acoustics. Most people with average hearing can fairly accurately tell the size of a room by its reverberant qualities. There can be no confusing the sound our voice makes in a gymnasium with that it makes in a small classroom. Because of the distance between the reflective surfaces in a gym, the sound of our voice takes a longer time to return to us than it does in the smaller classroom. We can therefore determine that the gym is the larger of the two rooms.

Another acoustical property of a room is whether or not it is "live" or "dead." A room is said to be live if it is lacking in absorbent materials. A room that is dead has an abundance of absorbent materials. This phenomenon can be readily observed when you paint a room, for instance. When the furniture, curtains, and rugs have been removed, the sound waves are unimpeded and are free to bounce from wall to wall. This reflective "echoey" sound may also be noticed in the shower and is somewhat responsible for the added "richness" of our otherwise average singing voices.

The acoustical quality of a studio begins in most cases on the drawing board of an acoustical designer. I say in most cases because one of the most sought-after recording studios in New York—Liederkrantz Hall—was once the home of an indoor beer garden. It was here that the New York Philharmonic Orchestra recorded much of its most celebrated music.

Perhaps it was the ornateness of the plaster walls, or the age of the walls, or the material in the plaster, or the thickness of the plaster; perhaps it was the wood, the marble, the height of the ceiling—whatever the reasons, it was an acoustically exquisite building.

Studios such as those found in Liederkrantz Hall are unusual; you will not always work in acoustically desirable situations. As a result, you learn to improvise.

An effective way of determining whether a room is lively or dead is by clapping your hands together. If the resultant "slap" sound has some ring or echo to it, the room is lively. If the sound is without echo, the room may be considered dead. By repeating this process as you walk around the room, you will discover that sections of the room vary in acoustical response.

THE WAY WE HEAR

The ability to hear frequencies in the range of 20 to 20,000 Hz is dependent upon your age, sex, and the physical condition of your auditory nerve endings and eardrum. Children are capable of hearing up to 20,000 Hz, and women in general can hear higher frequencies than men. The condition of your ears is dependent upon many things. The amount of noise you've been exposed to and for what length of time is very important. A short burst of loudness is not as damaging as loud noises over a longer period of time. Once damaged, auditory nerve endings are irreparable. A loss of hearing can be quite insidious, but it is also very permanent.

Amplitude and Loudness

When a source creates a sound of great energy, the peaks and troughs of the wave increase in *amplitude* (height or peak), but the wavelength remains the same (see Figure 2–4).

Humans are capable of hearing sounds between 0 decibels (dB) (the threshhold of hearing) and 130 dB (the threshhold of pain). This means

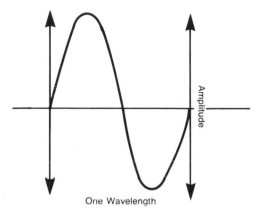

One Wavelength

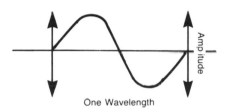

One Wavelength

FIGURE 2–4 The more energy in the vibrations, the greater the number of molecules that are compressed and rarefied. This increased movement of molecules affects the amplitude of the sound wave and increases the vibrations of the eardrum. This increased *amplitude* determines the way we perceive loudness.

that if you are at an airport and somehow are exposed to the engines of a jet plane (130 dB), the noise of the plane is 10,000,000,000,000 times greater than the threshold of hearing.

We arrive at that figure logarithmically, by taking the sound of the jet engine, 130 dB, which is 10 to the 13th power, or $10 \times 10 \times 10 \times 10 \times 10 \times 10 \times 10 \times 10 \times 10 \times 10 \times 10 \times 10 \times 10$, or 10,000,000,000,000 (ten trillion) times greater than the threshold of hearing. Incidentally, the threshold of hearing that we speak of does not mean the level where a sound can be discerned by all creatures on this earth, it only means a sound level where humans can detect the sound.

Volume Unit Meters

When we make the sound of a gunshot in front of a microphone, the *volume indicator* (needle) on the volume unit meter indicates the volume of the gunshot in terms of electrical energy, not acoustical energy (see Figure 2–5). The actual sound wave does not travel through conductors, but is converted into electrical energy.

When we speak of signals being "in the mud" or "in the red," we are referring to the levels that are in this area for any excessive length of time. Brief dips and peaks are unavoidable, but if they continue for any appreciable length of time, they can be controlled by "riding gain." This simply means bringing the levels up when they become too soft or reducing the levels when they become overly loud.

However, if you have to "reach" or "dig" (increase the volume excessively) to hear a low signal, you will also destroy the signal-to-noise ratio and you will amplify all the background and equipment noises from the studio.

Measurements of Hearing

We are accustomed to measurements having a linear fashion. Two inches is twice as long as one inch. Two pints is twice as much as one pint. Two times two equals four. The odometer and speedometer in our cars are linear. When we weigh our vegetables in the supermarket the scales are linear. If we double one pound we get two pounds. These all are examples of linear measurements. However, this is not how our ears hear things. Our hearing is not linear, but logarithmic.

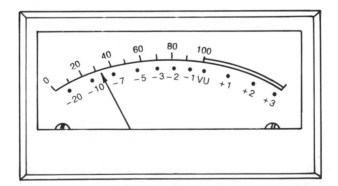

FIGURE 2–5 Pictured here is a volume unit meter used in broadcasting. The top row of numbers represents the percentage of modulation. This indicates the amount of signal passing through the meter in relationship to its capacity. Any signal below twenty dBs is too weak and is said to be "in the mud." Any signal above 100 Dbs is too strong and is considered "in the red." A continued loud signal will drive the needle off the scale, referred to as "pegging" the needle. Abuses such as these—if sustained—will cause the meter to be inaccurate. The bottom row of numbers is the volume unit scale.

Decibels

When we produce a sound or speak, we are changing the air pressure level about us. These changes in air pressure or sound pressure levels (SFL) are measured in terms of decibels. The decibel is actually a ratio between two sound intensities equal to ten times the common logarithm of this ratio. The logarithm of a number is that power to which ten must be raised to equal that number.

Sound Levels

Sound levels can be very deceiving when recording effects. If, for instance, you are recording subway sounds, you may have a train come roaring into the station with a meter reading of approximately 100 dB. If another train of equal loudness roars past at the same time as the first train, you would assume that the loudness level of the two trains would be doubled. Fortunately for us, this is not the case. When a sound level is "doubled" it is only increased by 3 dB. In order to actually double the noise level of a sound, it would have to made ten times louder.

The phenomenon we call "loudness" is a complicated one. This and other interesting aspects of sound are discussed in Chapter 3.

SUMMARY

1. The creation of sound effects is an art that requires both creative and technical skills.
2. Sound is caused by the vibration of molecules.
3. Without a medium of propagation or conductance, such as gases, liquids, or solids, humans would be unable to hear.
4. Sound will not travel in a vacuum.
5. Electrons are the negatively charged particles in the structure of the atom.
6. All liquids, gases, and solids are made up of atoms.
7. When a pool of water is disturbed, the waves move in an up and down fashion. These are called transverse waves.
8. When air is disturbed by sound, the resultant sound waves move away from the source of the disturbance. These are called longitudinal waves.
9. A sound disturbance causes molecules to become compressed in peaks and to draw apart in troughs (rarefaction).

10. An oscillation (vibration) of sound that causes one peak of molecules and its corresponding rarefaction is said to be one cycle of sound.

11. The amount of time a sound wave vibrates in one second is its frequency.

12. Frequency vibrations are measured in terms of hertz. A hertz is the number of cycles a sound wave completes in one second. If a sound vibrates at 100 cycles per second, its frequency is 100 Hz.

13. The distance between adjacent peaks of a sound wave is its wavelength.

14. When a sound comes in contact with a barrier it is either reflected, partially reflected, or totally reflected.

15. The velocity of sound is 1,100 feet per second.

16. The flow of electrons through a conductor is electricity. A conductor's capacity to accommodate the flow of electrons is dependent upon the elasticity of the conductor.

17. Echoes are caused by early reflections, while reverberations are caused by later reflections.

18. The more energy in the vibrations, the greater the number of molecules that are compressed in the sound wave.

19. Increasing the vibrations of the sound waves increases the movement of our eardrums and our perception of loudness.

20. The fundamental sound wave, when observed on an oscilloscope, looks like a single line rising in peaks and dipping to troughs. In actuality, a sound wave is a complex series of sound waves that vary in amplitude and wavelength and are superimposed on the fundamental sound wave.

21. Acoustics is the scientific study of sound reflection and absorption.

22. Humans have the ability to hear frequencies between 20 Hz and 20,000 Hz.

23. When a sound is produced, the sound pressure levels entering our ears changes, causing us to hear the sound.

24. Because loudness is such a subjective matter, we rely on meters for an accurate interpretation of loudness.

25. Sound is measured in terms of decibels.

C H A P T E R 3

WHAT IS A SOUND EFFECT?

The art of creating and using sound effects has become so sophisticated that sounds rarely appear in film or television as they are found at the source. With the emergence of sampled and synthesized sounds, every-one—not just the artist—has to have a thorough understanding of what goes into the makeup of a sound in order to either create a sound or to explicitly communicate about a sound. The days of critiquing a sound effect as "being too loud" or as "needing some highs" are gone. Today, when an artist is capable of building an effect from literally a single tone, it is vital that everyone knows exactly what constitutes a sound effect.

Every sound has its own distinctive wave form. That is what distinguishes a .357 Magnum gunshot from a toy cap pistol. And yet, in this highly complicated art of creating sound effects, other conditions apart from wave forms must be considered in order to successfully reproduce or create new sounds.

Listed below are the nine components that most influence how we perceive a sound effect. By modifying or eliminating any one or a combination of these components, you either slightly change the sound or create a totally new sound.

1. Pitch	**4.** Loudness	**6.** Sustain	**8.** Speed
2. Timbre	**5.** Attack	**7.** Decay	**9.** Rhythm
3. Harmonics			

PITCH

The pitch of a sound is determined by the frequency of the sound. However, when we hear a sound, we rarely describe it as such and such a number of frequencies. Normally, we refer to its pitch. *Pitch* refers to whether the sound is high (shrill) or low (bassy). Of course, what makes that sound seem high and shrill to our ears is the high number of sound waves (cycles) per second. Conversely, a low or bassy sound has fewer per second. Conversely, a low or bassy sound has fewer cycles per second.

Pitch also refers to the way we perceive frequency levels. Frequencies are normally grouped as low (bass), midrange, and high (treble).

COMPARISON OF PITCH AND CYCLES

Extreme Lows	Below 40 Hz
	-----BASS
Lows	40 to 300 Hz
Midrange	300 to 4000 Hz --MIDRANGE
Highs	4,000 to 10,000 Hz
	----TREBLE
Extreme Highs	10,000 to 20,000 Hz

Included in the low end of the scale are such sounds as thunder and gunshots; in the midrange, a telephone ringing; at the high end, small hand bells and cymbals.

Because our hearing depends on so many factors, determining the pitch of a sound is an entirely subjective matter. As we shall learn later, listening to a sound and hearing a sound are two entirely different matters.

A police whistle causes more vibrations than a clap of thunder; therefore, the pitch of the whistle is higher than that of the thunder. The piano string A above middle C vibrates at 440 Hz. Therefore, the A note is higher than the middle C note.

Continuing our piano analogy, a keyboard contains 88 keys and covers a frequency spectrum of 27 Hz to 4200 Hz. These various notes are broken down into divisions called *octaves*. An octave is the interval between any two frequencies that have a ratio of 2 : 1. Human beings are capable of hearing approximately ten octaves.

The first four octaves of the frequency spectrum compose the bass range; they sound powerful and warm. The fifth, sixth, and seventh octaves compose the midrange, the part of the frequency spectrum to which humans are most sensitive. This is also the frequency range that gives sounds their energy. The eighth octave gives a sound its "presence." Presence in a sound enables us to hear it clearly and gives us the feeling that we are close to its origin. The ninth and tenth octaves give a sound vital lifelike quality.

Utilizing Pitch

The pitch of a sound is its fundamental frequency without the presence of harmonics. Two examples of instruments capable of producing fundamental frequencies are the tuning fork and the pitch pipe. Although both of these instruments are capable of producing pure tones, the sound of a fundamental frequency without harmonics is dull and uninteresting.

When we listen to a tape of a fundamental frequency of 500 Hz, the sound is as uninteresting as that of the tuning fork or pitch pipe. Yet by applying what knowledge we have about the pitch of a sound, we can utilize the component pitch and create many new sounds.

The Doppler Effect

One of the many advantages of understanding how the various components of sound help us in the creation of sounds involves the "doppler" effect. First explained in 1842 by an Austrian physicist, Christian Johann Doppler, the doppler effect involves our perception of sounds regarding moving objects. The human ear collects sounds and directs them to the auditory canal. From there the sounds vibrate against the eardrum. These vibrations are transmitted from the middle to the inner ear, which contains fluid in which are immersed the auditory nerve endings. When vibrations disturb this fluid, the impulses are sent to the brain for interpretation. It is the condition of these nerve endings that determines the accuracy of the information that is transmitted to the brain.

The doppler effect states that in order to interpret sounds, the ear depends not only on the frequency of a sound that strikes the eardrum, but on the total *number* of sound waves that strike the eardrum. If the sound source remains stationary at a fixed distance, the frequency of the vibrations reaching the ear is the same as at the source. If, however, the sound source moves toward the listener, a greater number of sound waves strike the eardrum each second; therefore, the brain perceives the pitch as being higher.

If you were to stand on a platform as a train approached, the sound of the oncoming whistle would seem higher in pitch to you than it would to a person actually on the train. As the train continued past you, because there is a sudden drop in the number of sound waves entering the ear canal, the frequency pitch of the whistle would drop dramatically. Notice I didn't say the loudness level, but rather the frequency of the pitch.

This same phenomenon is experienced when you watch a parade. When the band is in the distance it has a "thin" sound; as it comes closer, it sounds "fuller" because of the increased number of sound waves. As it

passes, the sound once again thins out. The approaching, passing, and re-
ceding sounds of an automobile provide yet another example of this
phenomenon.

In none of the examples discussed has the loudness level increased
and then decreased; yet this is how this effect is most often achieved. The
proper way to create this effect is to filter out some of the low frequencies
as the train (or band, or car) approaches, to restore some of the lows as
the train nears, and then to make a sudden increase of low frequencies as
the train recedes.

TIMBRE

The timbre of a sound is that unique quality that sets it apart from all other
sounds. When you receive a phone call from a friend you haven't heard
from in years, your ability to recognize her voice is due to the timbre in
her voice. Even if you heard the voices of ten different women all reading
the same sentence, you would still be able to pick out hers. The same is
true of musical instruments. A listener can easily distinguish a trumpet,
piano, and violin all playing the same note because of each instrument's
distinctive timbre.

Timbre is that unique combination of fundamental frequency, harmonics, and overtones that gives each voice, musical instrument, and
sound effect its unique coloring and character.

Utilizing Timbre

Although manipulating a sound's pitch and timbre seems to be a Space
Age discovery, this technique was utilized back in the silent movie era by
musicians playing huge Wurlitzer organs. By pressing the proper combination of keys, these organists were capable of producing civil war battle
sounds with one hand and stirring music with the other. Although this
combination of sounds coming from one instrument favorably impressed
the audience, it wasn't as magical as most imagined. It was achieved with
the proper manipulation of the appropriate number of sound components,
not the least of which was timbre.

Today this "magic" is produced by sound generators. The results are
of such superior quality that even experts have difficulty distinguishing
natural and computerized music and sounds.

If a sound is made up of a fundamental frequency (the A above middle
C on a piano is 440 Hz), suppose we generated a fundamental tone of 440
Hz with something other than a piano and then repeated the fundamental
tone to form harmonics. Wouldn't the sound we created electronically be

that of the piano note A? The answer is yes. Although music purists complain that an electronic tone generator lacks the warmth (timbre) of a concert piano or a Stradivarius violin, the economics of the entertainment business strongly indicate that electronically produced music is here to stay.

HARMONICS

When an object vibrates it propagates sound waves of a certain frequency. This frequency, in turn, sets in motion frequency waves called harmonics.

Harmonics—or overtones, as they are sometimes called—are multiples of the basic frequency. There are two different types of harmonics—odd and even. The second harmonic of a frequency of 250 Hz is 500 Hz, the fifth harmonic is 1,250 Hz, and so on. Each harmonic that is added to a sound is somewhat weaker than the harmonic that preceded it. The combination of the fundamental frequency and its harmonics is a complex wave form. It is the formation of these complete wave forms that gives each musical instrument, and sound, its unique quality.

Again using our instrument analogy, the basic frequency and its resultant harmonics determine the timbre of a sound. The greater the number of harmonics, the more interesting is the sound that is produced. The production of the proper amount of harmonics is not automatic. A concert violinist and a novice playing the same violin will have drastically different results. The beginner will simply drag the bow across the strings with little regard for fingering or authority. As a result, the notes will sound uninteresting or even irritating to our ears. Conversely, the concert violinist will use his technique to produce a sound that is pleasing because of the number of harmonics produced. It is experience and technique that allow the concert musician the confidence to attack the strings with the appropriate strength to produce the maximum amount of harmonics that we find so satisfying.

We have discussed how poor technique with an instrument contributes to a poor sound, but what about an instrument that is either out of tune (not producing the proper pitch) or of such inferior quality that it makes it impossible to produce pleasing sounds? We have all heard the distinctive "pinging" sound that a fine crystal glass makes when its rim is struck. We also know that it is impossible to elicit the same sound from a glass that once contained peanut butter. The difference lies in the construction of the two glasses and their ability to vibrate when struck. The crystal glass has this ability, the inferior glass does not. It is therefore an object's ability to vibrate and set up harmonics that determines the pleasantness of the resultant tones.

Utilizing Harmonics

Manipulating the harmonics of a sound is a very useful device in comedy. By employing the technique of "dampening" (not allowing an object to vibrate naturally), many unexpected and comical effects can be achieved. Some door chimes, for instance, operate in concert to produce a pleasant little melody. This sound is most often identified with stately homes or mansions (at least in comedies.) If, however, upon pressing the doorbell of a beautiful home, one of the chimes is dampened so that in the middle of the melody a discordant "clunk" is heard, our opinion of the occupants becomes somewhat altered.

LOUDNESS

The loudness of a sound depends on the intensity of the sound stimulus. A dynamite explosion is louder than that of a cap pistol because of the greater amount of air molecules the dynamite is capable of displacing.

When a stimulus creates vibrations, air molecules are propagated in pressure waves. How many times these waves vibrate in one second determines their frequency, and the amount of air molecules compressed in a cycle determines the amplitude, or loudness, of a sound.

Utilizing Loudness

When we speak of something as being long or short, low or high, soft or loud, the words are meaningless unless we have a reference point. The measurement of a foot is long compared with an inch, but short compared with a yard. A play yard swing may be too high for the reach of a child but too low to accommodate an adult. The sound of a gunshot may be deafening in a small room, but actually go unnoticed if fired in a subway station when a train is roaring past. Loudness, as with everything else that is perceived, becomes meaningful only if we are able to compare it with something.

If we turn the level of a sound up by 3 dB, we have in effect doubled the level of loudness. If we turn up the level of a sound so we can readily perceive that it is "twice" as loud, we have actually made the sound ten times louder. These figures are based on the fact that we hear sounds logarithmically and not linearly. As you can see, the job of making a sound seem louder is more complicated than simply turning up the volume control fader. Perhaps the easiest solution is to introduce frequencies to which humans are most sensitive—those in the midrange between 250 Hz and 5,000 Hz.

If, for instance, we have two tone tapes of 100 Hz and 1,000 Hz and we set the loudness levels so that they both read 100 dB on the volume unit meter, our ears will readily perceive the 1000-Hz tone as being the loudest. Although the 100-Hz tone still reads the same level on the meter, we will have to turn up its fader in order to attain the same loudness level over the speaker as the 1,000-Hz tone. This phenomenon is called *equal loudness*. By knowing this, we can introduce high frequencies into the sound, either with the equalization pot or by layering (adding) another sound or sounds that have a great deal of midrange frequencies in them. This is why a clap of thunder in a horror movie may contain something so unweatherlike as a woman's scream.

ATTACK

An envelope of sound is composed of a sound's attack, sustain, and decay. The way a sound is initiated is called its attack. There are two types of attack: slow and fast.

In Figure 3–1, we see that the sound begins at A and reaches its peak level at B. At this point it drops slightly in level and remains steady until C. When the sound source is removed at point C, the sound decays to a point of silence at point D. By altering any one of these properties, we can change the sound.

Fast Attack

The closer the attack of a sound (A) is to the peak (B) of a sound, the faster its attack is. Such sounds as gunshots, slaps, and door slams are examples.

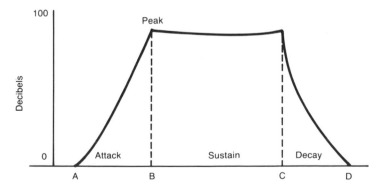

FIGURE 3–1   An envelope of sound.

Slow Attack

Sounds that have a slow attack take longer to build to the sustain level. A dog's short warning growl prior to a bark is one example. Stepping on a dried leaf, slowly tearing a sheet of paper, and closing a door slowly are some other sounds that have a slow attack.

Utilizing Attack

The suddeness of a sound achieving its sustain level contributes to the perceived loudness of the sound. Loud sounds are more frightening than soft sounds, and sudden loud sounds are the most frightening of all. If you are doing a scene about a woman alone in a house on a stormy night and you want to show how terrified she is of the situation, one way to accomplish this is by using loud claps of thunder. For maximum effect, edit any sounds prior to the peak level of the thunder. Although the sound hasn't been increased on the meter, it will seemingly be louder because of the suddeness of the attack (see Figure 3–2.)

Unfortunately, perhaps, the favorite time for directors to use thunder in films and television is when they show a shot of the window. Because lightning always precedes thunder, the suddenness of the sound is somewhat ameliorated by the flash of lightning. Therefore, to have this loud, sudden crash of thunder make an impact on the audience, we must add the element of surprise.

Suppose the woman decides to try to forget the storm by going to the den and getting herself a book to read. However, in reaching high overhead to a shelf, her hand accidentally knocks over a crystal figurine. As the anguished woman makes a frantic grab to catch it, it falls to the marble

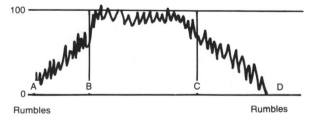

FIGURE 3–2 The entire thunderclap sound lasts from point A to point D, a time span of approximately 5 seconds. By starting the sound at point A, you have warned the audience of the impending thunderclap, and they are prepared for the sound. Because they have already heard the rumbling portion of the sound, the louder, more definitive part of the thunder is somewhat less effective.

floor and smashes. Only instead of hearing the familiar crash of glass, the audience hears a loud, sudden clap of thunder in its place.

This technique of using one sound to simulate a totally different sound can be extremely effective. However, in order for it to have the proper results, the audience must be surprised. If, for instance, we had used a thunder sound that had a slight rumbling prior to reaching its peak level, there would have been no correlation between that thunder sound and the fast attack of crashing glass.

SUSTAIN

Once a sound has reached its peak, the length of time that the sound will sustain is dependent upon the energy from the source vibrations. Once the source sound stops, the sound will begin to decay.

Manipulating the sustain time of a sound is yet another way of either modifying a sound or creating a totally new one.

Utilizing Sustain

One of the most important considerations given to a sound effect is its ability to be controlled. If, for instance, you are doing a film that takes place in New York City and you record the most realistic traffic sounds at 42nd Street and Broadway, what good are they if every time your actors speak, their lines are drowned out by realistic New York sirens? The answer for unmanageable sounds that must be heard in the background for any length of time is to *loop* the sounds. A loop of sound effects is a continuously running sound of something that can be stopped and started at any time without any noticeable change in the loudness level of the sound.

If, for instance, after you've taped your traffic sounds and you find that they are too "busy" (too many things going on at once), what you must do is find a portion of the tape where there are no extraneous sounds other than the ambiant roar of the traffic. By editing out the sound portion that sustains the type of sound you want and placing it on some type of cart machine (National Association of Broadcasters [NAB] or MacKenzie), you have an endless loop of sustained sounds without any other identifying changes such as attack, decay, or unwanted sounds such as sirens. If, however, you want to add sirens and horns, you have the control of those sounds on other carts, tapes, or film, and you can insert them where they won't interfere with your dialogue.

In the event you don't have a cart machine at your disposal, you must resort to the more time-consuming task of editing out all the unwanted sounds. Although a short length of tape can be spliced into a loop and played on a reel-to-reel tape machine, unless the tape is identical in content and loudness level, a loop of tape this short presents problems.

Suppose, for instance, we have the sound of an elevator starting, running, and stopping, all within 25 seconds, but the script calls for a scene to be played in an elevator that runs for 60 seconds. Again we edit the sustain portion and make it into a loop. However, because the running or sustained section of the sound is under 10 seconds, you must exercise tremendous care in editing, for unless the sound is without *glitches* (sudden bursts of level changes), a loop of sound this short will magnify these mistakes with a regularity that will soon become distracting—if not maddening—to the listener.

DECAY

The decrease in amplitude when a vibrating force has been removed is called *decay*. The actual time it takes for a sound to diminish to silence is the *decay time*. How gradual this sound decays is its *rate of decay*. If, for instance, you fire a pistol in a studio or a room that is extremely sound absorbent, there will be very little decay time, and more than likely what little there is will tail off quickly, giving the gunshot sound an unnatural popping quality.

The end of a sound is often referred to as the "tail" of a sound, and conversely, the beginning of a sound is its "head."

Whenever you are editing a sound, you must allow enough room at the tail for a natural fade. The fade may be extremely fast, but unless you are clipping a sound for a particular effect, there should be a fade.

This is no particular problem for a synthesizer, because it is done electronically. But what happens if you cut off the ending of an effect that is on tape? One method is to make another copy of the sound and this time, just prior to where the sound is clipped, fade the sound out fast. This will give you a sound with the proper fade on your second tape that, if need be, can be edited back onto the original tape.

Utilizing Decay

Although the decay of a sound is most influenced by the removal of its source, the resistance a sound encounters as it travels through the air also plays an important part in the rate of its decay. In the case of our gunshot, simply listening to the length of time it takes for the shot to decay tells us a great deal about where the shot was fired. If, for instance, the shot has a "tight" sound (little or no reverberation with very little decay), we know that it was fired in a small enclosed area with a great deal of absorbency. If the shot has a long decay coupled with an echo, we know that the shot was fired outdoors and its decay was most influenced by the resistance encountered by the air's molecular movement.

This knowledge is important when trying to simulate sounds. If, for instance, you have the sound of a shot that was fired indoors and you want to use it for a scene that takes place outdoors, by adding reverberation to the sound and "goosing" (quickly increasing the volume level) the attack of the gunshot momentarily, and by then doing a long slow fade, you have effectively created a sound most identified with an unconfined area.

SPEED

The measurable velocity of a sound determines its speed. By increasing or decreasing the speed of a sound, you can not only change the properties of one particular sound, but you have within your power to change the sound of an African waterfall into that of the detonation of the atom bomb!

Utilizing Speed

Audio tapes normally record and play back sounds at speeds of $7\frac{1}{2}$ inches per sound (IPS) or 15 IPS. If you record a sound at $7\frac{1}{2}$ IPS and play it back at 15 IPS, the information you recorded will be twice as fast as the recorded speed. If, for instance, you record an explosion at $7\frac{1}{2}$ IPS, it will sound like a gunshot when it is played back at 15 IPS. Conversely, if you record a gunshot at 15 IPS and play it back at $7\frac{1}{2}$ IPS, it will sound like an explosion.

To better understand this, if you were to play a tape of a 500-Hz tone that was recorded $7\frac{1}{2}$ IPS, and you played it back at $7\frac{1}{2}$ IPS, it would still sound like a 500-Hz tone. However, if you played this tape back at 15 IPS, the frequency of the tone would double and the tone's pitch would be perceived as higher. If you record a voice at $7\frac{1}{2}$ IPS and play it back at 15, the sound of the voice will not only be faster but higher, and it will resemble that of the cartoon chipmunk characters. (See Figure 3–3.)

As you can see in Figures 3–4 and 3–5, the letter "R," when recorded at a speed of 15 IPS, is exactly twice the length of the letter "R" recorded at $7\frac{1}{2}$. Although these two tape speeds serve music and dialogue well, sound effects needs a tape machine that is capable of playing a wide range of speeds. A piece of equipment with this capability is called a variable speed oscillator (VSO).

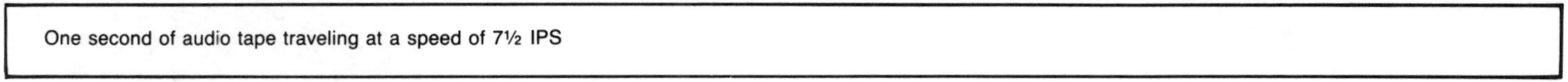

One second of audio tape traveling at a speed of 7½ IPS

FIGURE 3–3 One second of audio tape traveling at a speed of $7\frac{1}{2}$ IPS.

FIGURE 3–4 This is the letter "R" edited from the word *READY* at a tape speed of 15 IPS.

Variable Speed Oscillators

Working in conjunction with the reel-to-reel tape machine is the VSO. This piece of equipment is used to change the speed of the recorder's motor.

Most pieces of equipment have motors that operate on a standard 60-Hz power line. However, it is sometimes desirable to operate equipment at other than standard speeds. This is the function of the VSO. It substitutes this normal 60-Hz frequency with a variable frequency that is capable of controlling the speed of the motors in smooth increments, much as a dimmer on a light switch (see Figure 3–6).

RHYTHM

Rhythm is most often identified with music. In that context it is also associated with such terms as meter, cadence, and tempo. Therefore, rather than confuse the issue, for our purposes we will define rhythm as a recurring sound that alternates between strong and weak elements. Although there is a slight difference between rhythm, cadence, and tempo, the three will be grouped together under the one heading of rhythm, and distinctions will be made where necessary.

Utilizing Rhythm

In this highly complicated and subjective business of reproducing existing sounds and creating new ones, you will very often have to implement all the components that we have discussed in order to create the desired effect. Sometimes, however, one component will be so dominant that simply emphasizing that part of the sound will be sufficient.

For many years, the manual effect for marching feet was a group of wooden pegs suspended by wires in a wooden frame (see Figure 1–17).

FIGURE 3–5 This is the letter "R" edited from the word *READY* at a tape speed of $7\frac{1}{2}$ IPS.

FIGURE 3–6 A portable variable speed oscillator (VSO) attached to a tape machine. Most modern machines have the VSO built into the circuit. The VSO has another function that is extremely helpful in making difficult tape edits. By taking sounds or music and slowing them down with the VSO, it is often possible to hear spaces in modulation that will allow you to get in with a clear edit.

By properly simulating the cadence of marching feet, this sound effect was extremely believable. Of course, a group of wooden pegs would hardly suggest the precise sound of a large group of people marching in order. Rather, the rhythm supplied by the artist made this effect successful. If the rhythmic cadence were ignored, the wooden pegs would sound exactly like wooden pegs being drummed mechanically on a wooden surface.

UTILIZING ALL THE COMPONENTS

In the early 1950s at CBS in New York, because the film for the six o'clock news very often arrived at the network too late for sophisticated editing, the sound portion was frequently unacceptable for broadcast. Because this was a recurring problem, it was decided that rather than show the films without the sound, the various stories would be sweetened with sound effects.

On most occasions, this meant simply adding some type of ambient sounds to cover the edits. If, for instance, the story took place on the streets of a city, the artist would supply background traffic noises. And if the story involved a fire, the artist would furnish sirens and perhaps fire sounds.

Because of the time factor, the artists generally were not expected to be too specific with their sounds. However, occasionally they were called upon to develop sounds in a matter of minutes that should have taken days, such as the sounds to accompany footage of an atom bomb detonation. Although the picture portion of the telecast was duly impressive, the sound portion was less than spectacular. Not that the audio engineers didn't try. Located many miles from the site of detonation, microphones— unlike cameras with telephoto lenses—were unable to compensate for their distance from the subject. As a result, the sound portion of the explosion was little more than a rumbling, irritating, hissing noise with no explosive attack.

So disappointing was this sound that it became the reponsibility of the sound effects artist doing the news show that evening to come up with a sound more suitable for a news event of this importance.

In viewing the film, the artist was not only impressed with the magnitude of the explosion, but also with the seemingly deliberate amount of time it took for the mushroom-shaped cloud to take shape. Although the CBS sound effects library contained many large explosion sounds, none matched the awesome slowness of the picture he had just witnessed. Therefore, it was not so much the loudness or attack of the explosion that concerned the artist, but finding a sound that would match the inordinately long sustain and decay of the blast. Perhaps if the artist had had more time, the sound that was finally selected would have been different. But with air time only a matter of minutes away, the artist chose two explosion sounds (dynamite and a building being detonated) and the one sound he knew would give the atomic bomb a distinctive sound—a recording of the Mogambi Waterfalls.

It might surprise you that not one viewer, not one critic, not even the scientists who actually developed the bomb complained about this rather odd mixture of sounds being used for the atomic bomb explosion. And the reason was simple. What everyone heard that evening on the news matched so convincingly with the picture of the explosion that there was never any doubt in anyone's mind that the sound was not authentic. Yet if the sound of the explosion, no matter how realistic it was in nature, had been a shade too fast, it would have caused many viewers to think "something" was not quite right.

If you are impressed with the fact that anything so "unbomblike" as an African waterfall was selected from so many other more logical sounds, consider this. On that same news program, the Mogambi Waterfalls record was also used for the sound of a New York Times printing press!

In utilizing the components of a sound to create other sounds, the first step you must take is to disassociate the names of sounds with the sounds themselves. Although there are millions of names for sounds, the sounds themselves all fall into certain frequency parameters that can be manipulated by the nine components of a sound. Therefore, the fact that a waterfall record was selected is of no consequence. One sound of a waterfall is much the same as any other waterfall sound. The only thing we are interested in as far as sound effects are concerned is the magnitude of the waterfall's sound. Whether we use the Mogambi Waterfalls or the Niagara Waterfalls or even a "New York Subway Roaring through the 79th Street Station" is of little consequence; it is the roaring power sound we are after, not the exotic name.

Because these sounds offer no identification other than a constant roar, they can be readily adapted for other sounds. In the case of the atomic bomb explosion, the first consideration was to find a roaring sound that matched the speed of the picture. In order to do that, the turntable speed was altered from 78 RPM to approximately 30 RPM. At this slow speed, the roar of the waterfalls approximated the sustained portion of the explosion. Next, in order to have a good percussive attack, another record was selected. This time the sound was that of a building being detonated by a huge dynamite blast. This record was also slowed. When the cue came, the dynamite blast supplied the attack and the waterfall record supplied the sustain and decay.

Later, when there was a story that took place near the New York Times presses, the speed of the waterfall was increased and the loudness level was increased and decreased rhythmically to simulate the shot of the newspapers' machinery.

To do these effects on reel-to-reel audio tape, you need two machines, both of which must be equipped with VSOs to slow the speed of the waterfall and the dynamite explosion sounds. When the cue for the atomic bomb explosion comes, begin with the dynamite explosion for its good attack, and cross-fade (smoothly fade one sound in and the other sound out) the dynamite explosion out and the waterfall sound in.

If you only have one tape machine, the procedure is somewhat different. You will still need the variable speed capability on your machine, but now, since you are only playing the waterfall sound, you have to find a way for this one tape to supply the attack as well as the sustain and decay.

A strong percussive sound such as an explosion needs to begin at peak loudness level. There are two ways of achieving this, neither of which involves turning up the loudness fader; because no matter how fast you turn up the fader, the loudness level will have a slow, "faded in" attack sound. A better method is to edit the waterfall tape so there is blank leader at the head of the tape; therefore, when the cue comes, your sound is cued so that you immediately hear maximum loudness level with no attack time.

After the attack of the sound has been established, you sustain the loudness level until the picture indicates that much of the bomb's energy has been expended, then you begin a slow fade in the loudness level to simulate the decay of the sound.

Focusing On One Component

Very often, simply simulating one of the more distinctive qualities of a sound, such as its rhythm, is enough to convince the listeners that they are hearing the sound in total. An example of this is the cheers of a football game. For instance, you are in postproduction sweetening game highlights from the "big" game between Yale and Harvard and you suddenly discover a shot of cheerleaders on the Yale side of the field exhorting the crowd to cheer in sync with the flash cards, Y-A-L-E. Unfortunately, the game mikes didn't pick up the sound, and now you want to feature that shot in your opening. How would you fix it?

Unfortunately, this sound, like so many other distinctive sounds, is not readily available even at a commercial sound effects library. That is why it is so important to understand the makeup of sound so that you can create a facsimile sound that is as good as the source.

The first thing you must do is to discern the most outstanding characteristic of a crowd of people cheering in sync. In order to help you, let's try a little experiment. Whisper the letters: Y-A-L-E. If done with just the proper breath control, the hoarse indistinctiveness of your voice matches that of an overall crowd sound. Now whisper the letter "Y" and sustain the sound. With a little imagination you can create a fairly respectable crowd sound, and by varying the different components of this breath sound (and the amplification of this sound with a microphone), you can create literally hundreds of different sounds. This was the basic principle behind the Mogambi Waterfalls: to slow this sound to the point where it was basically a tone of indistinct noise. Now that you have accomplished this vocally, simply by giving your sustained breathy noise a series of sharp attacks in the rhythm of Y-A-L-E, you should, with a little practice, come up with a very respectable cheering sound without having to enunciate the letters Y-A-L-E.

This same procedure could be followed with any short crowd cheer you could steal from the film. The only restriction would be that the cheering sounds sustained long enough for the spelling of Y-A-L-E. And who knows, later in the game you might use this same crowd noise to cheer on that other venerable Ivy League school H-A-R-V-A-R-D.

In reproducing existing sounds or creating new ones, remember that the nine components of a sound are the building blocks for any sound effect. Whether you use the sound of the Mogambi Waterfalls for that of the atom bomb or that of a cheering crowd at a football game, concentrate on the sound of the effect and ignore its source.

SUMMARY

1. Our perception of a sound is influenced by one or more of the following components: pitch, timbre, harmonics, loudness, attack, sustain, decay, speed, and rhythm.

2. The pitch of a sound is determined by the frequency of the sound.

3. An octave is the interval between any two frequencies that have a ratio of 2 : 1.

4. Human beings are capable of hearing ten octaves.

5. The doppler effect depends not only on the frequency of a sound that strikes the eardrum but the number of sound waves that strike the eardrum.

6. The timbre of a sound is the unique quality that sets it apart from all other sounds.

7. Harmonics are multiple frequencies of the basic frequency.

8. The loudness of a sound depends on the intensity of the stimulus.

9. Doubling a sound's loudness increases it by only 3 dB. To actually make a sound twice as loud requires that the sound be made ten times louder.

10. The way a sound is initiated is its attack.

11. The sustain portion of a sound envelope is that period of time after a sound has reached maximum loudness level and before it begins to decay.

12. Loops of sound normally are the sustain portion of a sound envelope.

13. The actual time it takes a sound to diminish to silence is its decay time.

14. The measurable velocity of a sound determines its speed.

15. Changing the speed of a sound influences the pitch of a sound.

16. A VSO is used to govern the speed of a tape machine.

17. The rhythm of a sound is its recurring alternations between strong and weak elements.

18. Very often sounds are more identifiable by one of their components than by their sound alone. It is the manipulation of these components that makes it possible to use an African waterfall sound for that of an atomic bomb being detonated . . . and vice versa.

C H A P T E R   4

THE CATEGORIES OF
SOUND EFFECTS

We now have some basic knowledge about sound and the components that make up sound effects. Before we are ready to begin actually creating sound effects, we must know what function they will serve in the story. At this point, it makes no difference whether the effects are to be used in theater, television, radio, records, or film, because here all we are interested in is what category the effects will fall into.

With a deep apology to Gertrude Stein . . . a rose is a rose is a rose . . . but in sound effects, a hit on the head is not simply a hit on the head; the artist must know what *type* of hit on the head it is. Is it to sound natural, grotesque, fanciful, or slapstick?

If the creation of sound effects were an audio science, we would simply take a microphone and the appropriate recording equipment, go to the source, and record whatever natural sound we needed to satisfy the demands of the story. However, rare indeed is the director, or producer, who is entirely satisfied with all the sounds that are presented to him/her. No matter what the source or how natural and realistic the effects may sound, everyone wants something that will sound just a little "different."

To give you some idea of the subjectivity of sound effects, the following list of adjectives all describe one common sound effect. Take a moment and try to think of which one it might be.

Perfunctory	Tentative
Urgent	Bold
Desperate	Timid
Respectful	Angry

Hesitant	Persistent
Fearful	Hurried
Offhanded	Worried
Firm	Hysterical
Crisp	Authoritative
Rapid	Gentle
Thoughtful	Halting
Resolute	Faint
Aggressive	Strident
Fervent	Respectful
Measured	Inhibited
Exasperated	Apprehensive

All of these descriptions have been used on a countless number of shows to indicate how one sound effect is to be done—the *door knock!* If door knocks need to receive such special consideration, what about the countless other sounds?

Telling a decorator you want your kitchen painted "white" is of little help other than ruling out "red." But at least with paint, you can show the colors that you want. With sounds, the problem of communication is more difficult. As a result, before a "different" type of sound can be created, it must be established what category the sound will come under. Is the sound of the "hit on the head" to be used in a serious drama or a slapstick comedy along the lines of *The Three Stooges?*

There was a secretary at CBS named Margo Phelps. Margo was a delightful person and marvelous at all functions that came under the heading of secretarial; but her job in the sound effects department called for devotion above and beyond normal secretarial tasks.

When Margo first made the jump out of the secretarial pool, she was delighted. She was thrilled to be working with so many "creative and exciting people." Then she found out just why this particular job paid a little more than similar ones.

What follows is an example of a typical conversation between Margo and one of those "creative and exciting people."

> MARGO
> Oh, Bill, there's an added effect on your new
> show. They want the sound of a train.
>
> BILL
> What kind of train?
>
> MARGO
> They just said a train....
>
> BILL
> Running or standing still?

> MARGO
>
> ... a train. ...
>
> BILL
>
> Express or freight?
>
> MARGO
>
> ... all they said was ...
>
> BILL
>
> Fast or slow?
>
> MARGO
>
> I'll call them back.

Getting off the phone after a lengthy conversation, Margo began her recitation.

> MARGO
>
> It's a long freight train ... running on a level track for ten pages. It then climbs a long hill for fifteen pages and then comes to a stop in a station. It stays there for three pages and pulls out ... and, oh yes ... there are four train whistles ... two on level track, one going up the hill, and one when it pulls out of the station.

Margo, feeling she had covered any possible question Bill might possibly have, smiled confidently, "Is there something else you'd like to know about the train?"

> BILL
>
> Steam or diesel?

And just as it is important to know whether a train is steam or diesel, it is equally important to know whether this train appears in a World War II drama set in Germany or in a Disney cartoon. By understanding the various categories of sound effects, you will be able to better communicate and select the one sound effect that you want from the many sounds that are available.

NATURAL SOUNDS

Natural sounds are actual, unadorned sounds. Although popular in the early days of radio and film, today a sound's origin is not as important as the listener's expectation of how something *should* sound.

If an expedition from *National Geographic* travels to the northern polar regions to film the eating habits of polar bears, the sounds on the sound track will be those of the bear seen in the film. If there is anything disappointing about the sounds, it has nothing to do with the bear, only our level of expectations regarding the bear's behavior.

But this is a scientific study. Because the sounds that the bear makes are so important, the film is shot first, and the script is written later to comment on the recorded sounds—no matter how they sound—to help us better understand bears.

The use of natural sounds in a scientific study is understandable because a sufficient amount of time can be spent explaining a sound that might not meet the audience's expectations. Indeed, if two polar bears meet in confrontation over territorial rights and they both rear up to their 9-foot heights and merely whimper at one another, our expectations are disappointed: the whimpering sounds do not properly match the picture of two wild beasts with bared fangs. Again, because of the unusualness (in our opinion) of these sounds and behavior, much time can be spent explaining *why* this is normal deportment among bears. But how do natural sounds serve us in a less scientific environment?

On *Chico and the Man*, the NBC comedy starring Jack Albertson and Freddie Prince, the producer wanted to achieve an honesty reflected not only in the tools on the bench, but also in the sounds of the cars that came in and out of the garage.

After weeks of expensive studio time recording natural sounds from various cars, it became obvious that the results were not satisfactory. On one particular show involving a battered Pinto and Sammy Davis' personal Stutz Bearcat, hours were spent recording the two engines. But when the tapes were played back, the Pinto's engine sounded so much better than the Stutz Bearcar's that on the air the sounds of the two engines were switched. So much for honesty.

I might add this is not unusual. In films and in television, many natural sounds do not meet everyone's expectations. When this happens, they are either replaced with more suitable sounds or the natural sound is layered (other sounds are added) to make it more desirable.

The Shot Heard 'Round TV

Utilizing sounds, whether or not they are natural, that contradict an audience's expectations can sometimes have serious consequences.

During the live days of television, NBC in Hollywood did a production of *Petrified Forest*. Everything about the program was absolutely first-rate, except the sound of the gunshots supplied by Bobby Holmes, the sound effects artist.

Humphrey Bogart, making a rare appearance on live television, received excellent reviews, as did the rest of the production, save for the sound effects, which fared rather poorly. The next day, June 6, 1955, the

Hollywood Mirror News began a scathing review with this headline . . . "The Shot Heard 'Round TV."

The reason for all the furor was quite simple and familiar to the sound effects artist. Because of the magnitude of the star and other production values, little if any time was spent with something as insignificant as a few gunshots. The producer simply insisted the artist use authentic .38 pistols and dismissed the issue to direct his attention to the more pressing problems regarding good pictures.

What America heard that night was the actual, natural sound of a .38 pistol fired in the sound effects room of studio three. Unfortunately, the American audience was accustomed to hearing big, booming gunshots such as those characteristic of John Wayne westerns (and done in post production)! Of course, Humphrey Bogart's gunshots, despite their natural sound, came off sounding like an "anemic" cap pistol.

This is a serious decision that all artists and directors must make. Simply because you see the natural sound does not dismiss your responsibility for the resultant sound. If the sound you select, regardless of its authenticity, is too jarring to conventional concepts, you are inviting disbelief not only in that particular sound, but in other production values as well.

To further illustrate, the following excerpt from an article in *Electronic Magazine* explains very succinctly audience expectations and "opposing public opinion" regarding the sound of a gunshot ricochet.

> *The average person, however, has only heard ricochets through the medium of motion pictures. And a survey of several films indicates that only one type of ricochet has been used extensively —the more dramatic gradually decreasing pitch.*

Even as far back as radio, many directors insisted on the sound effects artist using the actual item for their sounds. Some of them carried it to rather extreme degrees.

The brilliant Orson Welles began his career in radio. On one occasion he was doing a story that took place in the Sahara Desert. Rather than having the sound effect artists do the subtle steps in the sand as they normally did, he decided to let the actors do their own steps. Welles arranged for a truck to deliver a huge load of sand to the studio. When Orson Welles did a desert story, he did a desert story! Although it looked rather impressive and convincing in the studio, watching the actors doing their own steps in the sand, the engineer in the booth found it impossible to achieve any sort of balance between the actors' steps and their voices. When the engineer made the microphone level loud enough to hear the steps, he raised the studio volume level so loud that he had no control over the dialogue. Although the engineer complained to Welles about this problem, Welles remained adamant about his decision to have the actors make their own sounds. However, later that day when everyone returned from lunch, no one in the cast and certainly neither of the sound effects artists was really surprised to find the sand in the studio gone.

Although this particular experiment with sound wasn't successful, it was Welles' intense interest in sound that would be so invaluable to his later film career.

Tampering With Reality

The problem with using natural sounds in a dramatic situation is the same one that silent actors faced when the talking picture came into vogue: living up to audience's expectations. This is a serious consideration. Since the advent of sound in film, it has become increasingly popular to tamper with the reality of natural production sounds. With the increased sophistication of audio equipment, the temptation to sweeten intensifies. When the hero bites into an apple is it "crunchy" enough for his macho image? When the heroine takes a shower, does the water sound "wet" and "sexy" enough? Do the ice cubes make enough "clink" . . . are the footsteps "forceful" . . . does the silk negligee "whisper"?

If directors continue their predilection toward tampering with the reality of sounds, we could see a generation of children who are disappointed with the sounds of their ice cubes, showers, apples, Fourth of July firework displays, picnics at the beach, walks in the woods, even their breakfast cereal. Although this sounds like a gross exaggeration simply to make a point, it wasn't too long ago that the Federal Communications Commission and the Truth In Advertising authorities banned the use of sound effects to enhance the sale of toy race cars, trucks, tanks, and other mechanical toys because of the disappointment suffered by children because their toys, "didn't make any noises like on television."

Although most problems with sweetening natural sounds is more a question of degree than legality, directors should exercise restraint and taste in postproduction. To help you with this, many years ago, before the accepted use of laugh tracks, a producer in New York decided to use "canned" applause. He felt justified for doing this because a terrible snowstorm had kept people from attending his popular game show. His advice to the sound effects artist who was playing the applause record was simply this, ". . . only play the applause where it would normally go if the storm hadn't kept our regular audience away. I don't want more or less, just what the audience at home is accustomed to hearing when they watch this program and it isn't snowing a blizzard."

Good honest advice. Unfortunately, the excuses for using applause records became more elaborate. "They're not applauding because . . . the political news is depressing them . . . the economic situation is bad . . .

they've been standing out in the line too long . . . Bloomingdales is having their annual white sale . . . Macys and Gimbels. . . ."

What started out as an excuse for a snowstorm snowballed. No longer were game shows affected by anything so unpredictable as audience attendance: producers had something better: a recording of people applauding for a show they were not attending. But if they *were* there, that's how a game show audience would sound . . . honestly.

Finally, to show how seriously our perception of reality has been altered in regard to sounds, a veteran of World War II gave this rather startling account of his first day in combat at Anzio. "Believe it or not, the first thing I thought about was [how] it didn't sound like a war. Having grown up watching Hollywood war movies, I expected a lot more sounds and much bigger sounds, it wasn't until I [was] hit that I realized what I was in was real."

CHARACTERISTIC SOUNDS

A natural sound becomes characteristic when it is either altered or imitated by other sounds to satisfy either personal dictates or production demands. Foremost among these demands is that sounds be recognizable and that they meet certain audience expectations. Yet, what are these expectations that audiences have, and how are they formed? What made the sounds of Sammy Davis' expensive automobile engine completely unacceptable? Weren't they the actual, natural sounds? Wouldn't it seem logical that if the sound of that particular engine met with the approval of the design engineers, it would certainly meet the standards of producers in show business? And just what are these "standards"?

Actually, there are no such things as standards. Most directors are slavish to the details of certain sounds, but when forced to choose between authenticity and dramatic impact, they will always opt for the latter and use "dramatic license" as an excuse.

If, of course, the same dramatic results can be provided by the natural sounds, so much the better; but this is rarely the case. Sound effects supply two important elements in drama and comedy shows—to inform and to appeal to some emotion. It is usually only when the sounds provide information that directors will allow artists the freedom to select natural sounds.

Sounds on Commercials

This business of natural versus characteristic sounds is such a serious one that car commercial producers avoid the risk of losing potential customers with what might prove to be disappointing engine sounds by using the less revealing and more emotionally manipulative music tracks. The only reason characteristic car tracks are not used is the truth in advertising law. Years ago in radio, before this ruling was enacted, all automobile sounds (excluding comedy effects), regardless of what they were supposed to be, were supplied by that of a 1935 Deusenberg! The smooth, throaty hum of this extraordinary motorcar became so identified with luxury that its motor sound was used extensively—even on commercials for other cars!

There were two excellent reasons for the selection of the Deusenberg car, neither of which had anything to do with excellence of engineering.

Low frequency sounds have a warm, reassuring sound. Perhaps this stems from our infancy when parents often make soft humming sounds or hummed lullabies. Hearing those same "humming" sounds in an automobile can reassure us that the car is safe and has a "nice" sound. The other reason, perhaps less romantic, is that the low frequency sound of the Deusenberg engine didn't get in the way of the actors' dialogue.

Pictures and Effects

Two of the major influences on our lives are what we see and what we hear. When the two are in harmony, the results are extremely satisfying. "Tall waves of white curling surf pounding on a sandy beach." "A chilling autumn's rain drumming against a window while we snuggle near a roaring fire." "The quietness of a winter's snowfall silencing the countryside." Sights and sounds—what a source of joy when they complement each other! But as we know, this ideal combination of picture and sound is indeed a rarity in the real world. Although film and television often portray reality, they are a form of art, and as with all art forms, they deal with reality with a fanciful attitude.

Sounds in the real world are not orchestrated to individual needs or moods. We may not like them, but we must accept them because that is the way of reality. In films and television, we create a reality that has a certain viewpoint. In real life we may be attending a funeral, devastated by the sadness of the moment, when suddenly in the distance, a car drives by with its radio loudly playing rock 'n' roll music. We may be angry at the thoughtlessness of the person, but we can't argue with the reality of the situation.

If we used those same sounds in a film we created, the only way they would be accepted is if the story was so strong and the production values so uncompromising that the audience looked upon the scene as reality and empathized with the characters for the intrusion upon their grief.

Sounds such as these, that suddenly change the normalcy and familiarity of a mood, test the believability of a scene to its utmost. If they work, they are extremely effective in counterpointing the scene's mood. When they don't, the sounds stand out as inappropriate and do nothing more than upstage the scene's action.

At this point it should be thoroughly understood that a visual scene creates its own mood—sounds merely support this mood. A bright sunny day isn't less bright or sunny because of the absence of birds singing. Yet when the two are together, we say the day is "perfect." Nothing is missing. In postproduction, the visual portion of the scene is already on film, all the sounds can do is support that scene. Whether sounds are natural or characteristic, there must be a reason for their being.

COMEDY SOUNDS

Perhaps the best definition of comedy was given by Jackie Gleason: "Pal . . . it's what you get a laugh with!" In sound effects, this translates into timing, the element of surprise, and the comical quality of the effect itself.

Timing is that particular instant when all conditions are working together to bring about the most desired results—laughs. The *element of surprise* is doing what the audience least expects in the most incongruous manner in order to cause laughs. The *comical quality* of an effect is what makes people laugh. So far, Jackie Gleason is right.

COMEDY STYLES

As far as sound effects are concerned, there are three different types of comedy styles: *straight, slapstick,* and *cartoon.*

Straight Sounds

Straight sounds in comedy can be either natural or characteristic. The one thing they can't be is funny. These sounds are used for such straight or noncomedic purposes as cueing, transitions, or backgrounds.

In burlesque, comedy was often used as a filler—something to fill time and give the stage crew an opportunity to change the sets of large production numbers. Because of this, comedy sketches were often done in front of the proscenium curtain without the benefit of a setting. To give the audience some idea of where the action in the sketch was taking place, they were given introductions that were both brief and to the point. Very often a comic would simply walk out on the stage, slam a rubber cactus plant down on the stage, and announce loudly to the audience, "Well, here we are in Mexico!" Of course, this same cactus was used for Spain, Brazil, Texas, Arizona, or any location that was simply hot. The point is, they never wasted time with anything that wasn't funny. Straight sounds are the burlesque cactus. They are only there if they serve some vital purpose.

If, for instance a comedy scene is played in a car, the sound of the car starting, driving off, and the engine running are all done with sounds that are the least distracting both in content and loudness level. Once the car is running under the scene it should be either played at a level so low that it is hardly heard or taken out completely under the first good laugh. Comedy depends on the playful receptiveness of the audience, and any sound effects that are too realistic intrude on this playfulness. As one comedy director put it, "I don't mind you playing a car sound behind the scene, I just don't want to hear it."

Slapstick Sounds

Charlie Chaplin once referred to humor as "playful pain" and pointed out that when a situation became overtragic it was funny. That definition of humor by Chaplin perhaps describes the vital function that sound effects have in slapstick comedy. The sounds are designed to assure that the audience accepts the characters' painful experiences as being "playful."

The term *slapstick* comes from a highly physical form of comedy that was done in vaudeville and burlesque. In those days, an assortment of props and drummer's traps were used to accent the numerous slaps, punches, and hits that the participants of this broad comedy form rained on one another. The most popular of these devices was a *slapstick*. Today, although there is considerably less personal mayhem and abuse in slapstick comedy, it still depends a great deal on tumult, pratfalls, props, and sound effects for its laughs.

The best examples of slapstick sounds can be heard in the old *Three Stooges* movies, although many of the older comedies such as those starring Laurel and Hardy, the Marx Brothers, and W. C. Fields also used outrageous sounds to complement their physical humor. Keep in mind that these films were made not too long after vaudeville and the silent films, and audiences were familiar with this violent brand of humor. Although the sounds were added to punctuate certain physical acts, the sounds also told the audience that these acts were totally without relation to reality and therefore should not be taken seriously.

Slapstick sounds work best when they emphasize pain. They help the fun by adding the element of "too" to the comedy. By making the sounds "too long," "too short," "too loud," "too big," they take the hurt out of the pain. If, for instance, a man gets his hand stepped on, the accompanying sound effect should exaggerate the reality of what might happen—broken bones in the hand—by being "too crunching and too loud and long" to be mistaken for reality. Because the sound is suggestive of what could really happen and because of the manner in which the comedian plays off the sound (reacts to it), the scene can be funny. Suppose that instead of selecting a sound that gives us too much of reality we go to the other extreme and select a sound such as a bulb horn, as was so often used in Three Stooges movies. The reason this worked for the stooges is that all their sight gags (physical humor) were done at a very fast pace. In this way, they never gave an audience time to think about what was being done. It was simply one fast physical piece of business after another in rapid succession. It was because of this rapid pace that the quality of sound effects was far less important than how quickly a sound, any sound, could accent three slaps, two nose tweaks, and one eye stab, without slowing the action. Not only do the sounds draw focus by punctuating the action, but they make the physical hittings and abusive actions less realistic and therefore less hurtful by matching them with ridiculous sounds. And that is one of the main functions of slapstick sounds, to connect a sound so absurd to a physical action that audience knows, "That hit can't possibly hurt . . . it sounds too funny."

Subtle Comedy

Although situations are the bases for most comedy on television, comedy that is derived from a person's character and the resultant attitude that a character has *about* a situation is subtle comedy.

The subtle humor of the radio show *Amos 'n' Andy* was so popular during the depression years of the 1930s that the country almost came to a standstill as millions of Americans tuned in to NBC to hear the ongoing warm and humorous story of Amos and his good friend Andy and all of the other beloved brotherhood members of "The Mystic Knights of the Sea." During the 15-minute version of this program, Freeman Gosden and Charles Correll played the majority of characters. So sensitive were they about anyone seeing them without blackface makeup or costumes that they requested that the sound effects artist select and deliver the effects to the studio, but not be present in the studio while the show was on the air. Because of the infrequent and simplistic sound effects needs of the show, the request was granted.

An example of the economy and taste that the gentlemen employed regarding sound effects is illustrated by their use of an ordinary intercom buzzer sound.

Andy always wanted things that were beyond his acumen and monetary resources. True to his character, when he opened a small office, he wanted it to be first-class. This, of course, meant acquiring a secretary and an impressive sounding interoffice buzzer. Although Andy was successful in talking his girlfriend into being his secretary at a rate of "no money to start, but more when the business gets on its feet," he was less successful at installing the buzzer. Unable to afford an electrician to do the job properly, he talked a fellow member of "The Mystic Knights of the Sea" into helping him. Unfortunately, the friend knew even less than Andy about the correct procedure for connecting an interoffice buzzer. As a result, they succeeded in hooking the buzzer up backwards.

This meant that every time that Andy needed his secretary, he had to go to the door and holler, "Miss Blue, would you buzz me?"

> MISS BLUE
> Yes, Mister Brown.

SOUND: SMALL BUZZER

> ANDY
> Now, come in here, please ... and, oh, bring your
> book, ah may think of something important....

For years this went on. A delightful comedy device employing an inexpensive "small buzzer" and a lovable character who saw nothing wrong with having to tell his secretary to buzz him each time he wanted to summon her to his office.

This is an excellent example of "less being more." Had anything been done to make the sound of the buzzer funny, it would have been too much. After all, Andy and his friend didn't make the buzzer; if they had, the emphasis would have been on the buzzer sound itself (the sound of which would reflect the two friends' ineptness, thus deserving the laugh). But inasmuch as they only wired the buzzer wrong, the only function the sound of the buzzer had was to imitate somewhat the sound of a legitimate intercom system and to act as a reminder as to how foolish we all sound when we try to be something we aren't.

CARTOON SOUNDS

Long before there was a Mickey Mouse or a Donald Duck, there was Punch and Judy; and when Judy hauled off and hit Punch, the puppeteers discovered their audiences enjoyed the violence more if they amplified the various hitting sounds. Sound familiar? Then what makes slapstick sounds different from cartoon sounds? Actually, very little—with one important exception. These same puppeteers also discovered that when Punch tried to retaliate and missed, his frustration and anger were heightened and therefore made funnier again by the addition of sound effects. This then is the difference between slapstick and cartoon sounds. Cartoon sounds emphasize not only action in the most absurd and amusing fashion, but they also allow the audience to hear a character's *emotions!*

This all began when cartoons were a rather crude affair. Because of the problems presented by poor sound quality, limited animation, and a need to tell the story as quickly as possible, sound effects were used to solve many of these problems. Not only did they furnish funny sounds for the action, but they helped eliminate some of the dialogue and animation by replacing them with sounds to indicate how the characters felt.

When a character in a cartoon became "boiling" mad or was hit on the head hard enough to hear "birds singing," this is what the audience heard. Although many of the effects used in those early films have been replaced with more modern sounds, the technique is still applicable today.

Picture and Sound Harmony

Although a harmony between picture and sound is important in all areas of sound effects, it is especially vital in animated films. Inasmuch as there are no natural sounds emanating from the cartoon characters, the mood of the film must be supplied either by music or sound effects.

In early cartoons, drummer's traps were used extensively for sound effects. When Popeye jumped from the mast of a ship to rescue his beloved Olive Oyle, his descent was followed by a slide whistle. When Mickey Mouse hit a baseball over the fence, the sound of the hit was provided by a wooden block, and the slide whistle sounded as the baseball went soaring into the air. Later in the inning when Donald Duck struck out, the motions of his missing the ball were highlighted by the sound of a wind whistle.

Those simplistic sounds worked because they were consistent and in harmony with the limited pen and ink animation. Even if the artists in those days had the means to use today's sound effects, it would have been a mistake, for there is no correlation between complex sounds and simplistic visual effects. The harmony of the film would be so disturbed that it would distract from the entertainment value of the cartoon itself.

Of course, modern audiences are accustomed to a big, loud stereophonic sound. So when they see the crafty Roadrunner being pursued by Wiley Coyote, the chase, tricks, and predicaments must be reflections of this big modern sound. Their's is the world of assorted frenetic actions, and with these come the appropriate sound effects. The quickness of the Roadrunner leaving a scene is emphasized with gunshot ricochets. Stops are highlighted with brake-smoking skids. And all explosions, despite their size, are atomic in loudness. In contrast, if you were to do a cartoon about Peter Rabbit, it doesn't matter how extensive your sound effects library is or how many synthesizers you have at your disposal—if the prevailing mood of the film is bucolic and innocent, you must select your sound effects as you do your music, to enhance that mood. Any variations, no matter how clever or imaginative they may be, will disrupt the harmony of the story. This is true of cartoons, slapstick comedy, fantasy, sitcoms, and soaps; wherever sound effects are used, always maintain the prevailing mood and never allow the sound effects to overshadow the scene.

FANTASY

Fantasy sounds deal with the world of the imagination. For our purposes, ethereal and science fiction sounds are included under the general heading of Fantasy. However, as you will see, there are differences that make them distinct.

Ethereal literally means "having the nature of ether; hence aerial or heavenly." Therefore, ethereal sounds are used to indicate movements or events of a heavenly nature. Prior to synthesizers and the multiple track tape machines, producers had basically two choices when it came to ethereal effects: music or a clap of thunder. Music indicated mystery and wonderment; thunder, revelation and anger. Although there have been a few notable exceptions, this stereotyped thinking exists even today.

Although *science fiction* sounds also fall into the category of fantasy, you are limited in your selection of sounds if you expect your audience to find them believable. Just as we've assigned thunder a God-like sound, science fiction must also contain characteristic sounds to which an audience can relate. If it doesn't, it will be rejected. Yet how can an audience be so demanding about a sound that no one has ever heard?

As we have discussed, sounds have certain properties that make them distinctive. We are normally frightened or alarmed by sounds that are short, loud, and explosive, such as gunshots or thunder. Therefore, even though you are dealing with fantasy and you want your sounds to be original, you can attain both these goals by using whatever sounds that you feel are appropriate as long as you satisfy an audience's general expectations as to what these sounds should be.

Believability and Expectations

When you think about it, audiences can be quite baffling. Even though we are dealing with fantasy, which by definition is not of this world, an audience will tolerate your being imaginative only with that with which they are already familiar.

Science fiction movie fans will never consent to any sounds as prosaic or mundane as a conventional gunshot; and yet, try to use a futuristic weapon capable of melting mountains that lacks these familiar loud explosive characteristics and the audience will question its validity.

This does not mean you can't use any sounds you want. It only means that if you stray too far from an audience's expectations, without some justification, you run the risk of losing them. The only justification an audience will accept is a well-written and believable story.

In real life we must believe the sounds that we hear no matter what our personal objections might be. A rifle with the firing power to pierce two inches of oak planking may make a short, small popping sound, but that is the reality of the weapon. As most of us have learned, there is little enough to be gained by arguing with the reality of a situation without arguing with the *sound* of a situation.

Films and television don't have the luxury of reason that reality offers. Therefore, films and television must concern themselves more with the believability of a situation because they don't have reality for an excuse.

Even if you are working on a fantasy or science fiction story, you can't fire a mountain-melting gun and produce an ineffectual sound unless you have an extraordinarily good reason. A reason, I might add, that is eminently clear to the audience. If you don't, it will simply cause too many questions in your audience's mind. Simply because what you are doing falls in the realm of fantasy isn't a good enough reason.

SUMMARY

1. All sound effects may be grouped into one of two categories: natural or characteristic. A natural sound is that of an actual source. A characteristic sound is what a sound should be according to someone's perception of the sound.

2. When a natural sound is manipulated in such a manner to achieve a desired effect, it becomes characteristic.

3. Inasmuch as we all perceive sounds in a personal manner, there are no standards regarding how an effect "should" sound.

4. Simply because you use the natural sound is no guarantee that it will be perceived as natural.

5. Our perception of natural sounds has become so influenced by ideology about sounds that we are often disappointed with reality.

6. Trying to match a few natural sounds is sometimes more time-consuming than redoing the tracks with characteristic sounds.

7. Whenever you have a time or budget problem, concentrate your efforts on a few selected scenes that are very obvious to the audience, and the sounds for the other scenes will be assumed by the audience.

8. If brevity is the soul of comedy, sound effects are its exclamation marks.

9. Sound effects ease the pain of physical humor.

10. Sound effects that draw attention to themselves have no place in comedy . . . or drama, for that matter.

11. Sound effects, wherever they are used, must reflect the mood of the story in the same manner that the music does.

12. The more skillfully written a story is, the less an audience will question the validity of sound effects.

13. Everyone has an idea of how something should sound. The problem facing the sound effects artist is to be creative without straying too far from these critical expectations.

CHAPTER 5

STUDIO AND POST SOUND EFFECTS EQUIPMENT

Although the actual sound effects for film and television are basically the same, the equipment and techniques used are somewhat different. Effects for films, other than those found on the production tracks, are done exclusively in post production; there are no effects done at the time of the filming. Conversely, sound effects for most nonfilmed shows in television are still being done in the studio. To accommodate the demands of this type of production, the techniques and equipment used in the studio are quite different than those used in post production.

Post production work takes place in a room filled with highly efficient, technologically advanced equipment. Here directors concentrate all their attention and efforts on sweetening the finished product. Of course, prior to this point was the studio work. It is in the studio that the director's attention is divided among dozens of problems, not the least of which are time schedules and budget demands. If for no other reason than these, expediency often wins over time-consuming artistic endeavor.

In order to meet production demands in a way that is both accommodating and economically feasible, studio equipment is somewhat less elaborate and sophisticated than that found in a large post production facility.

THE TELEVISION SOUND EFFECTS STUDIO

Since the introduction of digital sound, cart machines, and sound computers, equipment has changed at such an alarming rate that the television networks are hesitant to expend huge sums of money for equipment that will be outdated in a matter of months. As a result, the equipment found in the television studio is not the equipment found in the networks' post production rooms.

This decision not to become involved in the equipment race is not based entirely on budget considerations, but also on the precept: "If it isn't broken, don't fix it." Also, what sense does it make to have equipment with capabilities to perform functions that are quite beyond the needs and demands of the average television studio show?

The Needs of the Studio Show

Film has never attempted to add music and effects at the time of the shooting, and the trend in prime-time television is to follow film's post production operation. Producers of daytime television, because of their more limited time and budget allocations, still try to do as much of the show in the studio as possible. Although live network shows are almost exclusively sports and news programs, with the advent of video tape, a production procedure has been created to emulate live coverage and yet have the safety of tape. It is called "live-on-tape."

Live-On-Tape

Shows are called live-on-tape because they are done as if they were live, and barring a catastrophe, there are no stops. Talk shows use this format to elicit high energy performances from their guests. Game shows use it for somewhat the same reason, but perhaps more importantly, the game show producers like the control it offers. Ever since the game show scandals of the 1950s, networks have been extremely sensitive to any insinuation of hanky-panky regarding their game shows. As a result, a watchdog legal department (compliances and practices) monitors every show carefully as it is taking place. If there are no problems, the show runs as if it were live; if something occurs that is legally questionable, or for that matter morally offensive (bad language, indecent gesture), the program is stopped and the tape is edited. But as far as sound effects are concerned, every show must be treated as if it were live. This means using appropriate equipment.

Talk Shows

Other than the *Tonight Show* or *Late Night With David Letterman* sound effects on talk shows are almost nonexistent. In the case of these two shows, the effects are done in the studio as the show is being taped; they are done either manually or on some type of cart machine. The only difference between this type of show and an edited taped show is that if you make a mistake with the effects, you can stay up and listen to them that night, because under no circumstances will they be erased, altered, or edited out.

Game Show Sound Effects

Sound effects for game shows fall into one of three categories: (1) hard-wired, (2) closures, and (3) hands-on.

Hard-wired effects. Some production companies make provisions for sound effects in the actual design of their set. These effects are electronic and are hard-wired to perform some function relating to the game. If, for instance, it is important to have a different sound for each of ten numbers that flash rapidly on and off, it is far easier to wire the sounds into the same circuit as the lights so they both work together than it is to have a sound effects artist try to manually sync the effects with the lights.

Closures. A closure circuit is simply a method of starting an effect in sync with the action of the contestant. It is no more complicated than a light switch in your home that "closes" a circuit to turn on your light.

A closure differs from a hard-wired effect in that it is not a part of the set, but simply a method of playing a sound effect by remote control. These sound effects are in the equipment operated by the artist, but instead of the artist playing the effect, they are electronically started and stopped by some action taken by the contestants. If, for instance, contestants are required to push a button as soon as they know the answer to a question, normally these buttons are part of a "lock-out" circuit. When one button is pushed, the other is prevented from operating. To emphasize which one responded first, many shows have a particular effect for each contestant. Therefore, when contestant "A" pushes her button, a bell sounds; when contestant "B" pushes his button, a buzzer sounds. Both of these sounds are loaded in the sound effects cart machine, but both are actuated by the closure circuit.

Hands-on. Although manual effects performed into an open mike are now rarely or never done, most game shows use taped effects operated by the sound effects artist. Normally, these effects are done in cart machines to allow the greatest number of different effects. Typical game show effects include sounds for the following: big wins, loser, right answer, wrong answer, bankrupt, vowels, consonants, and whatever else the producer feels needs a "ping-bing-bong" or "buzzer."

Some shows even expand the artist's duties by supplying sounds for the prizes . . . "We'll fly you to Acapulco!" and on showing a film clip of the sponsoring airline's plane, the artist supplies the sound of a jet engine.

Game shows effects are usually predictable in that they use the same effects in a fairly repetitous manner. For this reason, the only demands are that the equipment perform reliably and that it is versatile enough to play effects either separately or in concert. For this, any multiple channel cart machine would be acceptable. To invest in equipment more sophisticated would be like using a Rolls Royce to deliver newspapers. Additionally, because the electronic circuitry used on modern game shows is so elaborate, malfunctions so frequently, and is so time-consuming to repair, the last thing a producer wants is to have to stop the show because of an electronic breakdown in the "bing-bong."

Sitcoms

Most situation comedies use very few sound effects. Those that are called for are usually the "doorbell-door knock-phone bell" variety, all of which can be done manually or on cart in the studio by the artist. Therefore, whether you do them there at the time of the taping or post them later is more a matter of personal preference and budget than of technical superiority.

One school of thought prefers to do the effects in the studio so the actors can react to the actual sounds. Very often, especially if the effect is something out of the ordinary such as the sound of someone falling down the stairs, the actor has something to react to and can find a piece of business or a "take" (facial reaction) that wouldn't normally be discovered if the effect had not been heard.

Another consideration is the audience. If right in the middle of a sketch a doorbell is supposed to ring and all they hear is a stage manager calling out "doorbell ring" or "ding-dong," it is bound to interrupt the audience's attention. In the live days of television comedy, interrupting an audience's attention during a sketch was tantamount to comedic disaster. It is all well and good to announce prior to doing a scene that the effects will be added later, but why encumber an audience with your technical problems when their whole purpose for being there is to give support to the cast by laughing.

When fans of a particular situation comedy come to the studio, they fully expect to see the show exactly as they are accustomed to seeing it at home. To dampen this mood of eager anticipation with too many production and technical problems is asking for trouble. One problem for an audience is watching the actors reacting to sound effects they're not hearing; their attention begins to wander and they feel cheated that what they are seeing isn't the finished product. How much these diversions affect the sense of play that is so basic to laughter is problematic, but there are many producers who feel the cost of doing the sound effects in the studio so that actors and audiences can hear them is too small to take a chance.

One producer of an extremely successful comedy show summed up her feelings this way:

> *These are the people that put our show in the top ten. I am grateful to them, and anything that I can do to make their visit to our studio pleasant, I will do. Now, if 250 people go away happy and they tell two people, and those 500 people tell two more people . . .*

On the other hand, many producers feel it is a waste of production money to pay a sound effects artist to be in the studio to perform a few effects that can be done just as easily in post production. These producers reason that audiences enjoy having a look "backstage," which home viewers never see. And as far as "losing" the audience because of production and technical inconveniences, they reason, isn't that what laugh tracks and post production are for?

Sitcoms without an Audience

If a comedy show is done without an audience, it is a whole different matter. Inasmuch as you don't have to be concerned with tape stoppages for technical or performance purposes, it is simply a matter of personal preference and budget considerations as to whether you do the effects in the studio or post production.

Variety Shows

There are basically two types of variety shows. One hires talent who do their own acts, and one has its regular cast of talent performing new material each week. The number of sound effects each of these shows uses is dependent on whether the show is performed live, live-on-tape, or stop and go, as are many of the sitcoms.

On variety shows that use the sitcom technique, much of the material done in the studio is performed with little or no rehearsal, and ad-libbing by the actors is encouraged. As a result, sound effects done in the studio are used mostly for cueing purposes, and the effects are "sweetened" in post production. The old *Laugh-In* show used this format. However, on a live show such as *Saturday Night Live* or a live-on-tape show such as the *Tonight Show,* everything is done the way you see it on the air, often with very little rehearsal. As a result, the equipment used on these shows must be adaptable.

To give you some idea of the amount of sound effects this can involve, "The Shooting Gallery" is a pantomime I wrote for Red Skelton (see pages 93–95). Although Red mimed all the props, sound effects played a very important role in helping the audience visualize the action. Effects for this script were done by Ross Murray.

The condition of the script purposely has not been changed so that you can get some idea of what a sound effect script looks like after several rehearsals. Unfortunately, during the actual taping of the show for "air," these scripts are very often more of a hindrance than a help, because in looking at the script you know what is supposed to happen, but Skelton is only interested in getting laughs from the studio audience. If, however, what he does on stage corresponds with what is supposed to happen in the script, fine; if not, and you want to keep your job, that's fine too.

This refusal to "stay on script" was not a conscious desire on the part of Skelton to make the sound effects artist's life more stressful, it was simply the way he felt most comfortable working. Jackie Gleason also refused to follow the script if he could get laughs without it. Both of these great comedians were products of vaudeville, where the ability to ad-lib according to the temperament of a particular crowd was necessary for success. Indeed, both had such successful careers prior to television, they saw no reason to change now. As a result, the sound effects artists who did the Skelton show relied heavily on the versatility of manual effects and multiple cart machines, because they knew they had to be ready for any of the last-second ad-libbing for which Red Skelton was so famous.

Once Ray Erlenborn, one of Skelton's regular sound effects artists, was caught totally unaware by one of Skelton's ad-libs, and the director demanded to know what happened. Ray shrugged his shoulders in an apologetic manner and said, "It was totally my fault, but I was working under a terrific handicap . . . I had a script."

Well aware of Skelton's fondness for sound effects as well as his inclination to ad-lib, Ray relied heavily on manual effects (see Figure 5–1). On hand were a Chinese gong, a stack of berry baskets to produce the sound of a door being smashed down, a comedy crash tub, and a wooden temple block. In addition to these live effects, Ray kept his fingers on a NAB cart machine control panel, thus giving him access to such taped sounds as "the little boidy chirping."

So much for the well-rehearsed, precise, and orderly world of the television studio. To accommodate this freewheeling and often chaotic

THE SHOOTING GALLERY

One thing I like about going to the
amuesment park is the shooting gallery.
May I have my hat please?

SOUND:MERRY-GO-ROUND MUSIC

RED IS WIDE-EYED AND A CHILD AGAIN.HE GAWKS AT THE
BRIGHT LIGHTS AND WATCHES WITH HIS HEAD MOVES AS:

SOUND: ROLLER COASTING ROARING BY *— Long cut!*

SOUND: MERRY-GO-ROUND MUSIC UP *Follow Red's look!*

RED'S HEAD BOBS UP AND DOWN AS HE WATCHES THE
PAINTED PONIES AND LAUGHING CHILDREN. STOPS AND
ORDERS SOME POP CORN. DIGS COIN OUT OF POCKET AND
PAYS. WHILE HE'S WAITING FOR ORDER,HE WAVES HAPPILY
TO THE CHILDREN ON THE RIDE. TAKES ORDER WITHOUT
LOOKING AT IT. STARTS WALKING,STILL LOOKING AT
THE WONDROUS SIGHTS. WITH TIPS OF FINGERS HE TAKES
SEVERAL SCOOPS OF "POP CORN." ON THE THIRD ONE...HE
FINALLY LOOKS AT "POP CORN" AND GETS A DISTASTEFUL
LOOK ON FACE. WIPES FINGERS ON SIDE OF PANTS AND
DOES WHAT HE SHOULD HAVE DONE IN THE FIRST PLACE...
LICK THE ICE CREAM CONE THE MAN GAVE HIM BY MISTAKE!
RED STILL AWED BY THE SIGHTS SOON FORGETS HIS LITTLE
MISTAKE AND HAPPILY LICKS CONE.

(Continues, next page)

method of doing sound effects in the television studio, the equipment must
be reliable and must lend itself to spur of the moment changes. Although
better equipment with far more sophisticated capabilities exists, the sound
effects done in the studio are only a part of a very complex production;
and any equipment that requires special consideration from an already

SOUND: FADE IN SHOOTING GALLERY

He'll get finger caught — wait for biz — cork pop.

RED WATCHES A MOMENT AND THEN GOES OVER A GIVES COIN
TO ATTENDANT FOR RIFLE. RED PUMPS RIFLE, AIMS AND
PULLS TRIGGER.

SOUND: SHOT

RED COCKS EAR AND WAITS TO HEAR BELL...NO SOUND. RED
DOES SGT. YORK BIT. WETS THUMB AND WIPES OFF GUN
SIGHT. AIMS AGAIN AND PULLS TRIGGER.

SOUND: SHOT AND BELL

THAT'S MORE LIKE IT! AIMS AGAIN...PULLS TRIGGER.

SOUND: SHOT AND BELL
AIMS AGAIN...PULLS TRIGGER. *change to 3*
SOUND: SHOT AND BELL *Fast Bells*

NOW RED IS GETTING COCKY! SHOOTS ONE HANDED.

SOUND: SHOT AND BELL

TAKES OUT MIRROR AND WIPES IT ON SHOULDER. TURNS BACK
TO TARGET AND FIRES LOOKING INTO MIRROR OVER HIS
SHOULDER.

SOUND: SHOT AND BELL

BENDS OVER AND SHOOTS WITH BACK TO TARGET THROUGH
LEGS.

SOUND: SHOT AND BELL.
Does fast draw bit
LOOKS AROUND SELF-CONSCIOUSLY AND THEN PRETENDS HE
IS RIDING A HORSE. SHOOTS AT THE BAD GUYS.

SOUND: SHOTS AND BELLS

SOUND: SHOT AND BELL

Sound: SHot 2and Bell

"The Shooting Gallery" (continued)

~~SHOT AND BELL~~

Fast →

TAKES AIM....ONLY THIS TIME...

SOUND: ~~NO SHOT~~ JUST BELL —

RED DOES TAKE AND LOOKS DOWN BARREL RIFLE... *Wait for*

SOUND: SHOT! *Red to squint*

RED DOES TAKE AND JAWS AT RIFLE. NOW RED DECIDES TO

DO SOMETHING REALLY BIG! HE AIMS RIFLE HIGH TO

THE LEFT.PULLS TRIGGER AND "WATCHES" THE PATH OF

THE BULLET. *In sync with Reds moves !*

SOUND: SHOT BELL RICCOCHET...BELL-RICCOCHET...

BELL-RICCOCHET...BELL-RICCOCHET.... *No Bell*

RED COCKS HIS EAR FOR THE SOUND OF THE BELL AND

REACTS TO BULLET IN REAR END! —— *Bass drum Crescendo*

TURNS TO CROWD AND MOUTH,"YOU AIN'T SEEN NOTHING

YET! PULLS OUT HANDKERCHIEF AND BLINDFOLDS HIMSELF.

PULLS TRIGGER *(wait — Red may do, cork pop*

SOUND: SHOT....... *Bit - Watch Him)*

RED PUT HAND TO EAR...NO BELL. FINALLY,TAKES OFF

BLINDFOLD...LOOKS AT TARGETS...AND THEN DOWN. DOES

TAKE. PLACES HANDKERCHIEF OVER FACE OF THE PRONE

AND UNFORTUNATE ATTENDANT...TIPS HIS HAT AND MAKES *hat over heart.*

A FAST EXIT AS WE.... *Sound: Merry Go Round — Up and into*

GO TO BLACK.

MUSIC: PLAYOFF

"The Shooting Gallery" (continued)

FIGURE 5–1 "SOUND: TEMPLE BLOCK WHEN RED SKELTON GETS HITS ON THE HEAD, FOLLOWED BY A LITTLE 'BOIDY' CHIRPING." As you can see, Ray Erlenborn must constantly divide his attention between the written script, which indicates what Skelton is supposed to be doing, and the monitor to see what Skelton is actually doing. (Photograph courtesy of Ray Erlenborn.)

harassed producer will be ignored. In the studio, when the stage manager asks for quiet and begins the countdown for taping, that means everyone had better be ready . . . or else.

THE SOUND EFFECTS CONSOLE

The console is the nerve center of the studio sound effects room. All the effects that are produced by the artist must first go through the console (see Figure 5–2). At this point it should be made clear that the output signal that leaves the sound effects room does not go directly to either the transmitter, as in the case of a live show, or to video tape recording, as in the case of taped shows. Instead, it goes to the audio booth to be mixed with the music and dialogue. Although on many shows (soaps in particular) this is not the final mix, the outputs from the various dialogue mikes, the music, and the sound effects are put on individual tracks by the audio mixer in the studio. The final audio mix of music, dialogue, and sound effects is done in postproduction.

MICROPHONES

One of the most important links in the sound chain is the microphone. Without it, there would be no sound on film, tape, or records. There would be no live radio or television, nor would there be the telephone. With all the considerations that are given to sound after it has been converted from

FIGURE 5–2 The purpose of the sound effects console is to centralize the various pieces of equipment. The output levels of microphones, turntables, and tape machines—both reel-to-reel and cart—are all different. In order to achieve consistency in the output level from the console to the audio booth, each piece of equipment has to go through the sound effects console. This is the amount of equipment a console must accommodate for the NBC daytime soap, *Days of Our Lives*. (Photograph courtesy of NBC.)

acoustical energy, you would think that at least as much attention would be given to the instrument responsible for the conversion—the microphone.

One of the few things all microphones have in common is that they are traducers that convert acoustical energy to electrical energy. Here the similarity ends.

Each type of microphone (and there are many) is designed to perform a particular function. One mike might attenuate the highs in a singing voice while another will boost the lows; yet another will emphasize the bass quality in the voice when placed close (proximity effect) or add more presence to the voice when moved farther away. Knowing what aspect of a sound you want to feature therefore is imperative when selecting the proper mike.

Microphone Characteristics

There are three basic polar patterns for microphones: omnidirectional, bidirectional, and unidirectional (cardioid). The patterns indicate the microphones' areas of sensitivity. The most sensitive area is referred to as "on axis" or the "live" side of the mike. The area that is least sensitive is said to be "off axis" or the "dead" side of the mike.

Omnidirectional microphones. An omnidirectional microphone picks up sound in a circular fashion, that is, it is sensitive to sound not only from the front part of the mike but from the sides as well.

When silent films gave way to "talkies," the omnidirectional mike was the only microphone available. Because it was so sensitive to sounds from all sides, it picked up camera noises and coughs as well as actors' dialogues.

Bidirectional microphones. The bidirectional microphone is also called the figure-eight microphone because its polar, or response, pattern resembles the numeral "8." The advantage of this mike over the omnidirectional mike is that two people can face each other and speak without having noise from either side *leaking* (being heard) into the mike.

Unidirectional microphones. The unidirectional microphone is most sensitive to sounds that are coming from a frontal direction. Because the polar pattern most resembles a heart, the mike is also called a cardioid mike. These mikes are especially popular with people performing in front of an audience.

The Sound Effect Microphone

Although a sound effect artist will use any type of a microphone that will get the job done, the condenser microphone with hypercardioid characteristics is generally preferred for both studio and field use (see Figure 5–3).

As you can see in Figure 5–3, the hypercardioid microphone is most sensitive to sounds that occur on axis, or to the front of the microphone at 0 degrees, and less sensitive to sounds that occur to the sides or rear. Although this is helpful in excluding many unwanted ambient sounds, very often sound effects artists must nevertheless travel to remote areas to obtain the silence they need to record sounds. Therefore, the next time you see a high-speed car chase through city streets on film or television, there is a very good possibility the car sounds were recorded in a remote area to give the artist the control over the car sounds without interference from other sounds. In Hollywood, many artists do their recordings in the Mojave Desert.

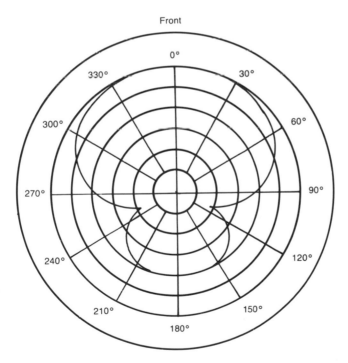

FIGURE 5–3 Diagram showing sensitivity pattern of a hypercardioid microphone.

Listening and Hearing

The microphone has often been compared with our own auditory system. Although this might seem logical, it isn't quite true. Our hearing is very selective; the elements of a microphone are not. Whereas humans can hear a number of noises at the same time but focus on only one, the microphone pays equal attention to all sounds.

The fact that humans do not hear all that microphones do can present a serious problem, especially if the effects are done "live." In a sense, humans hear what they want to hear. This ability to focus on certain sounds to the exclusion of others is what makes living in a noisy environment somewhat bearable. Can you imagine what it would be like if our ears gave all sounds equal consideration? This is what a microphone does. During a recording, sounds that are not obvious to our hearing are picked up by the microphone with the same noise-free fidelity that we want only with the effect we are trying to record.

Unfortunately, there is no true solution to this problem. You must simply be aware of all the sounds around you, use the microphone best suited for your needs, and follow the procedure used by veteran artists: keep recording a sound from different angles and at different levels until you achieve the sound you are looking for. Even if there were one correct method to record a sound, it is doubtful that anyone would be satisfied with the results.

REEL-TO-REEL TAPE MACHINES

Although cart machines have become extremely popular for performing sound effects, the reel-to-reel tape machine is indispensable for recording and editing new information. In addition to supplying sounds that will be transferred to carts, there are certain effects that can be done successfully only on a reel-to-reel machine. If, for instance you have a 4-minute scene that takes place in a moving car, rather than have the sound of the engine running on a cart, if the effect is on reel-to-reel and you have variable speed oscillator either as a function of the machine or as a separate piece of equipment connected to the tape machine, you can vary the speed of the engine simply by turning a dial. This will either slow down or speed up the tape machine's capstan motor, making the sound of the car's engine go slower or faster. Although you could prerecord changes in the car's speed and put the effects on carts, you would have no control over when the changes occur.

Analog Recording

Tape machines are the workhorses in sound effects and come in a variety of shapes and forms, each designed to perform a specific function. In addition, tape machines now come in two distinctly different formats: *analog* and *digital*.

When an electrical current is applied to a magnet, it sets up a magnetic field called a flux. In the case of the analog tape machine, the electrical current is analogous to, or representative of, the audio signal. When an audio tape passes the tape machine's record head, this magnetic flux causes the ferrous particles on the audio tape to become magnetized in accordance with the flux of the applied current. Unfortunately, an analog tape recorder will record any system noises as faithfully as it records the audio signal.

Digital Recording

Noise has been the enemy of audio since Edison invented the tubular wax phonograph record. Digital recording not only ends the noise problem, but it increases the dynamic range (a measurement between the lowest and loudest sounds) from approximately 75 dB with analog recording to 96 dB with digital. Although this increases the fidelity of what the digital recorder is capable of reproducing by eliminating all system noises, it makes any studio noises seem that much louder and more obvious, because those sounds once masked by system noises such as tape hiss are now clear and painfully obvious.

Digital recording, unlike analog recording, does not depend on the physical quantity relationship and electrical energy, but only on whether electrical energy is present. Therefore, when an analog wave is sampled, that sample is assigned certain binary digits called bits (*bi*nary digi*ts*). These bits, when grouped with other bits, are called words. Thus, a 1 or a 0 is a bit, and 101 is a word. All this means for our purposes is that the analog wave is converted into a stream of data words that are not affected by system noises, and then converted back to an analog wave at the output of the recorder.

To illustrate the difference between analog and digital recordings, consider a telephone conversation. When we receive a telephone call we have no difficulty understanding what the caller has to say as long as we have a clean line of transmission. If, however, there is a great deal of breakup and static, hearing that person becomes a great deal more difficult. Now suppose you and your friends live in an area where this is a constant problem; you might decide to use a system other than your voice to communicate about vital matters. If you each decided to communicate using Morse code, your problem would be solved because of the penetrating ability of the shrill dots and dashes. Furthermore, if you and your friends used this dot and dash system exclusively, "outsiders" unfamiliar with your system would be unable to decipher your messages.

Likewise, the digital recording system—the "outsider"—simply does not recognize equipment noise as communicable sound. By sampling a sound into a stream of data bits, it has effectively changed our telephone caller voices into a series of dots and dashes that is immune to system noises.

Analog Versus Digital

Studio sound effects either are accompanied by other sound effects or are heard under music, dialogue, or some type of ambience. The presence of these other sounds is usually enough to mask whatever slight noises might exist on the analog tape; therefore, the sounds do not present too serious a problem. However, when layering sounds on an analog machine, there is a great deal of dubbing from one tape to another, and each time a tape is dubbed, there is a generation loss in quality. This loss of generation, or sound clarity, is less desirable than slight tape noise. But because analog tapes can be edited so readily with equipment no more sophisticated than a razor blade, the analog tape recorder remains very popular with most studio sound effects artists. Additionally, if the artist is using a cart machine such as the MacKenzie, the need for analog tape is critical. Although a digital tape machine converts tape to analog, many artists believe this step is unnecessary and that it negates the advantages of the original noise-free digital sound. This, however, is true only with studio sound effect work in television. Most film sound effects artists now use the digital recorder exclusively for gathering sounds in the field.

Multitrack Tape Machines

If you, the sound effects artist, had the power to tell a producer how to produce, the director how to direct, and the actors how to act, then the only piece of electronic equipment you would need in the studio would be one multitrack reel-to-reel tape machine. And because the show would run flawlessly, all the sound effect cues would run sequentially. And if a scene required thunder, rain, and gunshots, all happening at once, the thunder would be on one track, the rain on a second, and the gunshots on a third. If the show needed more sounds, with a twenty-four-track tape machine you could play twenty-four sounds at once. Each would have its own time code and the art of sound effects would be no more complicated than laying sounds down on tracks and pushing a stop and start button. Just like the music business: kick drum on track 4; snare drum on track 7; high-hat cymbals on track 8; electric bass on track 3—each sound on a separate track time coded for ease of operation. Now go back and read the Red Skelton sketch. As you can see, there is nothing sequential about this sketch, and you are not dealing with time codes, but with visual cues from a comedian trying desperately to get laughs from an audience.

Although multitrack tape machines have the capability to produce a number of sounds at one time, because these tracks are on one tape, they don't have the independence or versatility for random selection. In post production, these machines are excellent, but for studio work, the artist must have a machine that offers immediate and numerous choices.

CART MACHINES

The Mackenzie

To better understand cart machines and their capabilities, let's examine the MacKenzie. Figure 5–4 shows an exposed view of a five-unit deck. On the far right are two toggle switches. The upper one is the power switch for the preamps, and the bottom switch controls the power to the motors. To the left of the motor switch is the master output gain control. The fader determines the overall loudness of the five channels.

To the far left are the trays holding the individual carts. To the right of the cart trays are the preamps. On the preamps are three controls: the fader on the left is for low or bass frequencies, the fader in the center is for high or treble frequencies, and the fader on the right is for volume control for that particular preamp.

To the right of the five-deck unit is the control panel. At the top is a volume unit meter that indicates the output levels from that particular deck, either from a single channel or a mix of all five channels. Beneath the meter is a sliding fader that controls the output gain of all five faders. Beneath the master fader are the individual faders that control the levels for each of the channels. Finally, at the bottom of the control panel are the stop and start switches for each of the five channels.

Figure 5–5 illustrates the heart of the MacKenzie cart machine. Each cart is capable of holding as little as 2 seconds' worth of lubricated magnetic $\frac{1}{4}$-inch tape and as much as 7 minutes' worth of tape played at 7.5 IPS. Each cart tray is driven by a common capstan and a single motor. One of the most important features of this machine is its instantaneous stop-start operation actuated by a high-speed solenoid.

FIGURE 5–4 The MacKenzie cart machine. (Photograph courtesy of MacKenzie Laboratories, Inc.)

These five channels make up a basic unit. Whether they are used in this manner or stacked, they are capable of producing a tremendous number of varied sounds.

The cueing system. Each of the five channels is equipped with photoelectric cells that operate in conjunction with transistorized pulse amplifiers that control the individual solenoids. A cue mark consisting of a narrow strip of highly reflective tape placed 3/4 of an inch from the beginning of the desired cue will give a start close enough to the manufacturer's "instantaneous" to make it acceptable for syncing such difficult cues as face punches. When this tape, with the reflective cue mark, passes one of the photoelectric cells, the resulting light pulse triggers the solenoid and the tape stops immediately and your effect is recued.

NAB Cart Machines

One of the alternatives to the MacKenzie machine is the NAB cart machine (see Figure 5–6). The NAB defines a cartridge as "a plastic or metal enclosure containing an endless loop of lubricated tape, wound on

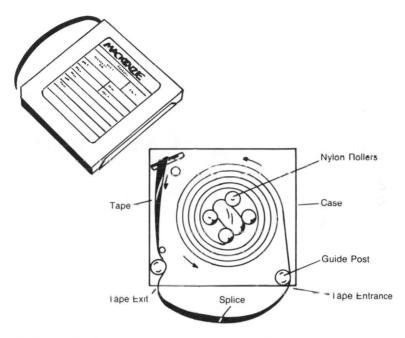

FIGURE 5–5 The heart of the MacKenzie cart machine.

a [rotary] hub in such a fashion as to allow continuous motion.'' Cartridges from various manufacturers differ slightly in details, but all cartridges used in NAB-standardized systems fit the preceding definition.

Similar to the MacKenzie, the cartridge tape consists of a synthetic base material approximately 1 mil (0.001 inch) thick. One side of the base is coated with ferric oxide particles for magnetic recording. The other surface is coated with a graphite layer for smooth operation. (See Figure 5–7.)

Transferring sounds to NAB carts. Transferring sound information to carts that have an NAB format is quite different. Here the tapes are already on the carts in varying time lengths. Always select a cart that is slightly longer than your material. Once that is done, unless you are absolutely certain the tape on your cart is clean, you must degausse (demagnetize) the tape, because NAB cart machines are not equipped with erase heads. Therefore, even in the record mode, information on an NAB cart will not erase.

When you are certain your tape is clean, you are ready to record. At this point, it is critical to start recording with the cart machine just prior to playing back the tape machine. If you don't, there will be some dead

FIGURE 5–6 A close-up of an NAB cart machine. (Photograph courtesy of Broadcast Electronics, Inc.)

air before the modulation starts. By depressing the record button (or stop buttons with some machines), a tone is put on the tape. After the tape has played, the playback head will "read" this tone and the machine will stop. Newer models include a cueing function that, once the tape has ended, fast forwards ($22\frac{1}{2}$ IPS) to the head of the tape and recues itself. This fast forward option is also available at anytime you want to rapidly recue the cart manually.

Some Similarities

The similarities between the NAB cart system and the MacKenzie are numerous. Both utilize an endless loop of tape, start and stop within 0.1 second maximum, run at speeds of 7.5 IPS, and are accurate within 0.1%. Furthermore, both types can be played singly or cascaded together to give greater mixing capabilities. There are other common features, but these are most important to the production of sound effects.

FIGURE 5–7 Both the MacKenzie and the NAB cart machines can be stacked to add greater selection of effects.

There are two significant differences. The first is the cueing system. As discussed earlier, the MacKenzie utilizes a photoelectric concept that is sensitive to the silver foil stop mark, while the NAB cart is cued electronically by a 1-kHz stop tone. Both are extremely efficient; however, the NAB tone system has an advantage over the MacKenzie in that the MacKenzie cart can build up shiny spots on the tape due to wear and fool the photoelectric cell into thinking it is the silver foil stop mark; thus the cart will stop before it is recycled. The other major advantage of the NAB cart is its fast forward capability (22.5 IPS). However, this feature is most useful with music carts, since extended loops of background sounds rarely if ever have a starting point that would need recueing, while most other sound effect cues are short enough not to need this fast forward feature.

WHAT ELSE HAVE YOU GOT?!

"I hate that effect . . . what else have you got?!" This is probably the statement most feared by sound effects artists. Although the post production effects artist isn't immune to such frantic requests, it is the studio artist who is most vulnerable. In the studio, usually during rehearsal, while the director waits nervously to hear "what else you've got," so too do a studio full of actors and technicians.

It would be nice if directors would take a few moments before rehearsal to preview, if not all, certainly the few effects that might be troublesome, but they rarely do. Even when they take the time to preview the effects, many directors have difficulty making an assessment of a sound without the picture it is to support. Because of this, very often excellent sounds are rejected because the directors lack the ability to "pictorialize" the sounds they are listening to. As a result, the artist must always be prepared for last-minute changes. The best way to prepare is to use multistacked cart machines, with backup sounds ready for all of your effects—and in the case of effects that might be troublesome, a number of backup choices. Effects are troublesome only in that they are judged subjectively. It isn't that the effects are more difficult to do or have a more unusual sound, they are simply subject to the director's educated whim. Two troublesome sounds are gunshots and punches. On soap operas that have alternate directors, for example, the artist will use a particular punch because that's what Monday's director wants; but when the artist uses that punch the next day, Tuesday's director hates it. And if the artist wants to keep working on the show, she won't say, "But Monday's director loved it." To simplify matters, use a cart machine that allows numerous effects to be played together or independently.

Another decided advantage of the cart machine is its ability to *layer* effects. If, for instance the script calls for a car crash, a cart machine allows you to use five channels with different aspects of the crash on five different carts. In this manner, each effect contributes to the overall sound. Perhaps on one channel there is a metal crash that has an excellent attack for the impact of the crash. On the second cart there might be a glass crash to assist the impact sound. On the third cart you might have a good metal wrenching sound that will sustain the crash, and on the fourth cart, a metal sound of a higher pitch that blends well with the third cart. The fifth cart might feature a metal crash sound that offers a good decay. Separately, they make a poor sounding crash; however, layered together, each furnishes a vital ingredient for the overall success of the crash sound.

The combination of wide selection and layering make the cart machine an important part of the television studio artist's equipment.

EQUIPMENT FOR POST PRODUCTION

Post production! Its that magical kingdom where all the mistakes you made on location or in the studio are corrected. The place where you finally have to make decisions about all those problems you couldn't or wouldn't decide on during the heat of the taping or filming. Post production! The room where you finally see what you really have in a show, rather that what you think and hope you have. Post production! The great fixing place—or is it?

Post production is literally the room where all the hard work you have done in the studio or on location is assembled. Once the envied domain of films, television has invaded these hallowed grounds to the point where films are more and more emulating many of television's editing techniques.

Often the difference between a pleasant and unpleasant experience in post production depends on your knowledge of what can and cannot be accomplished there. No matter how sophisticated the equipment in post production becomes, it will never displace good sound production habits in the studio or field. If you come prepared with all the ingredients you need, it is an exciting process to watch your project come together into a smooth finished product. If for whatever reason there are important ingredients missing, it can be a nerve-wrenching experience that takes a toll not only in time and money, but also on professional reputations.

Effects Editing for Film

Sound editing for film is done in one of two ways. Either the editor makes a splice in the film itself, sometimes called "sprocket editing," or the film is transferred to video tape and edited electronically.

On today's theatrical films, sound normally is divided into three categories: dialogue, music, and effects. In the final mix, it is the dialogue editor who, along with the director, has the final say.

The work atmospheres of film's sound effects editors and television's sound effects artists differ significantly. In television, all of the live and "live-on-tape" shows (the *Tonight Show* and *Late Night with David Letterman*) and most of the taped shows (soaps and game shows) have the sound effects artist in the studio itself, allowing direct communication between the artist and the director regarding cues or changes. In film, the editor is given a sound report and a reel of film and is expected to furnish the effects for that reel of film in a reasonable length of time with little or no communication with the director. While one editor is working on one reel, other editors assigned to that particular film are editing additional reels.

The Sound Cue Sheet

Cue sheets inform the editor of where the various effects go and their length. Listed at the top of the sheet are the numbers of the different tracks (reels of film) and the sounds that are to be found on each track. The numbers on the sides refer to film footage numbers. Head (beginning) and tail (end) footages indicate the length of the effects. An effect that has only a head cue means the sound is very short, usually under ten frames. The reason for all the various tracks is to give the sound editor maximum control over the selection of the effects, the levels of the effects, and their duration.

Some of the effects for the final film cut may be edited out and saved from the original sound production track, but in most cases, it is more expedient to simply use library effects. This is especially true of all but the most unusual background sounds.

A reel of 35-mm film is 1,000 feet long. Because of the large number of reels it takes to make a theatrical film and the complexities of the effects, there may be as many as ten effects editors working on the same picture at the same time, and this does not include the Foley artists.

When an editor is given a reel of film, he or she is also given a sound report as to what sounds are needed for that particular reel. Whereas this report will request "exterior sounds," the specifics of those sounds are left up to the discretion of the editor. This does not mean the editor has the last word. All sounds are subject to final approval by either the dialogue mixer or the director during the final mix. If the editor has selected a reel of crickets that get in the way of the dialogue, they will either be replaced by less abrasive insects or simply killed altogether. Or if the effects editor makes up a car crash effect and the composer has scored music for the crash, the director will make a choice as to what he or she wants featured. Rarely, if ever, are both given equal consideration.

Each reel of film is broken down into frames, sixteen per foot. The first twelve feet compose the "academy leader." This footage allows time for the projector to get up to speed. Therefore, all effect notations begin with 12 or a larger number. The number 12 would indicate that the effect begins immediately after the academy leader, while the number 12 + 10 means the effect occurs at the 12-foot and ten frame mark.

Figure 5–8 illustrates four tracks of effects for a scene that lasts for 32 seconds. All of the effects start immediately at the number 12, or at the expiration of the academy leader. On track number 2 we have assigned crickets. As you can see, they start at the number 12 and continue until the end at the 32-foot mark. The same is true of the wind sounds on track number 1. On track number 3, because the camera is on a wide shot, we don't start sneaking in the crow effects until the 18-foot mark on the counter; again, the effects stay in until the end of the scene. On track number 4, the laughter doesn't come in until the 25-foot mark and continues only until the counter reaches the 30 feet and 9 frame mark. Because both the picture and four effect tracks all began at zero, the effects and picture are in perfect sync.

The Movieola

In sprocket editing, the picture portion of a film is viewed and heard over a small projector called a *movieola*. This device enables the editor to view the film frame by frame and hear the sound in the same manner. If,

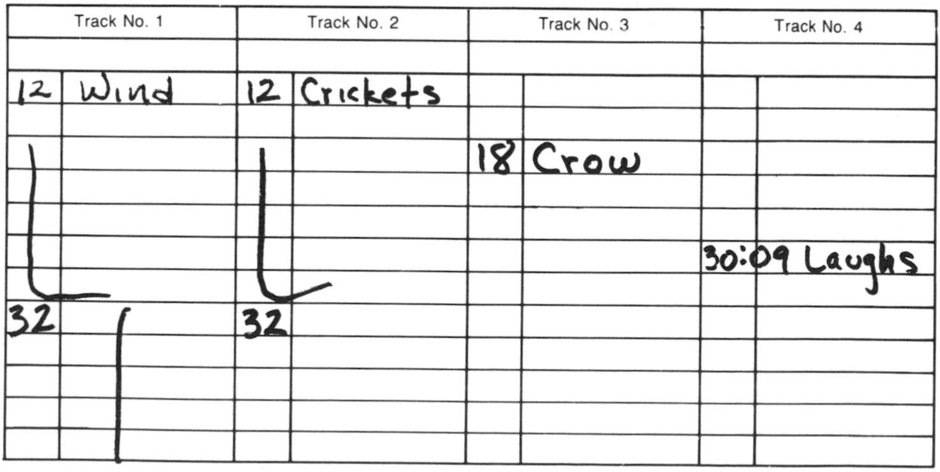

Track No. 1	Track No. 2	Track No. 3	Track No. 4
12 Wind	12 Crickets		
		18 Crow	
			30:09 Laughs
32	32		

FIGURE 5–8 Four tracks of effects for a scene lasting 32 seconds.

for instance, a door close is edited in at 28 + 3, the editor will mark with a grease pencil a line across the film at that precise measurement and initial what the effect is, so that the mixer will know when to open the pot rather than trying to keep track of footage and frames for effects. It takes a mixer approximately three frames to open and close a pot.

Film Synchronizers

In television, video and audio tapes are kept in sync by an audio signal, and information can be found in a matter of moments by the Society of Motion Picture and Television Engineers (SMPTE) time code. In film, synchronization between two or more reels of film is accomplished by the sprocket holes along the sides of the film. These perforations keep the film locked in place, and when the film is played, a counter ticks off the exact amount of footage and/or frames. These synchronizers usually come in sets of four. This enables the editor to add effects to four different reels at the same time.

Video Editing

Perhaps the innovation in television that was most responsible for the success of post production was the development of a sophisticated system for editing videotape electronically. To do this, an arrangement was needed that could locate various cues precisely and store this information.

Such a system was devised in 1970 by the SMPTE. The SMPTE time code assigns each video frame a unique number that can be read by any other video or audio recorder equipped with this system. The time code standard as adopted by SMPTE is an 80-bit digital code defining the hour, minute, second, and frame number of that particular segment of tape.

Thus the director has an accurate accounting of everything that was laid down on tape, including wild shots, pickups, and scenes shot silent. When this tape is played back in post production in conjunction with computer editors, it is possible to quickly and accurately locate a scene or shot that needs sweetening with sound effects. If for some reason the sweetening does not meet with everyone's approval, it can be done over and over until everyone is happy.

If, for instance, a scene requires the sound of a gunshot, the assistant director refers to her production log and sees that it occurs at 10 : 09 : 25 : 09. The videotape operator rolls the tape to that number and is ready to record the gunshot.

This can be done in one of two ways. The artist can "eyeball" the effect, either by using the business cue itself or the more accurate visual time code number shown on the screen. Alternatively, if the cue is more complicated, the sound can be transferred to a reel-to-reel audio tape machine that is in sync with the videotape machine, thereby accurately matching the sound and picture down to the last frame.

These time codes are "time of day" configurations. Simply stated, this means a 24-hour clock system is utilized to denote the times. Therefore, by using the more accurate number 13 to denote one o'clock in the afternoon, the confusion between 1:00 A.M. and 1:00 P.M. is avoided. This would appear as 13 : 03 : 21 : 16. The first two numbers indicate the hour; the second two numbers, the minutes; the third two numbers, the seconds; and the final two numbers, the frames or thirtieths of a second.

Film Versus Video

If ever there was a classic example of comparing apples and oranges, it is comparing the merits of film editing versus video editing. How can you compare the editing practices of a film that is expected to gross hundreds of millions of dollars with a film made for television? How can you compare the editing practices of the number one prime-time network television program with that of a modest nature film made for cable? Therefore, there is no "best" way of editing, merely a question of time limitations and budget.

SUMMARY

1. Studio sound effects are laid down in one of two ways: either as the show is being done or in post production.

2. Studio or field equipment is far less sophisticated than the equipment used in post production. However, studio equipment more than makes up for this lack of sophistication by being more adaptable.

3. All network sound effects rooms are equipped with a sound effects console, reel-to-reel tape machines, turntables, cart machines, and microphones for manual effects.

4. The sound effects console is the nerve center of all the sound effects equipment found in the studio. Each piece of equipment has its signal equalized, processed, and mixed by the console before it is sent to the audio mixer for the final program mix.

5. The three general pickup patterns for microphones are omnidirectional, bidirectional, and unidirectional. The pattern most used in sound effects work is the cardioid or heart-shaped pattern.

6. The signal from an analog tape machine is an electrical representation of the sound pressure's wavelength.

7. Digital tape machines sample portions of a sound wave and convert it into a series of pulses.

8. Reel-to-reel tape machines are advantageous due to the amount of material they can store and the ease and quickness with which this material can be located and edited.

9. Reel-to-reel tape machines equipped with a VSO allow the artist to vary the speed and pitch of music and effects.

10. There are basically two types of cart machines. The MacKenzie depends on a metalized tape and a photoelectric cell for cueing purposes, while NAB cart machines are cued by a burst of tone that is placed on the tape.

11. The advantages of cart machines include their adaptability to change and the control they afford the sound effects artist.

12. Layering sounds entails mixing a number of different sounds together to achieve another more complete sound.

13. The advantage of using multitrack tapes or carts is that each individual sound can be changed without disturbing the rest of the mix.

14. When looping sounds on a cart, a good rule to remember is that the more identifiable the sounds are, the longer the loop should be.

15. Because NAB cart machines are not equipped with erase heads, care must be taken to see the tapes are either clean or degaussed (erased) before laying down new information.

16. The primary difference between doing sound effects in the studio and in post production is the emphasis given to the two areas. In the studio, effects are only a part of the production. In post production, more time and attention can be paid to the effects because the two most important aspects of the production, pictures and dialogue, have been completed.

17. Because post production is interested only in assembling and editing creative work that has already been completed, the equipment involved is geared for technical perfection rather than the vagaries of the human element.

18. The major difference between film and tape sound effects editing is the technique used to synchronize the effects.

19. In film, the editing technique utilizes sprocket holes; in video, the SMPTE time code is used.

CREATING SOUND EFFECTS

CREATING SOUND EFFECTS

One of the most difficult jobs sound effects artists have is to throw away a sound effect—any sound effect—because they live in fear that some day that may be the very sound they so desperately need. As a result, there are literally millions of sounds carefully catalogued at television and film libraries around the world, waiting to be heard. Some of the sounds haven't been used since the days of radio and early television and films. But they all have one thing in common, someone, somewhere, at one time thought they were unique enough to save. After all, how can you possibly discard a sound so unusual as: "Heart beats. Cut 1: good valve. Cut 2: bad valve." "King Kong roars—cage rattle in bg (background)." "*Gone With The Wind* Atlanta fire—good crackle—lots of timber crashes." "Seig heils for Hitler—some goose stepping marching." "London air raid—sirens and explosions." "Ma Perkin's rocking chair squeak." "Buck Rogers' disintegrator gun—add some highs." "Midway Island aerial battle—plenty of good ack-ack—watch for the scream at the end."

Simply reading the pithy descriptions of the various sounds is fascinating. Although many of the sounds in these libraries were created by artists for specific situations or a director's particular taste, many of the effects were recorded at the source. Yet with all of these millions of available sounds, there isn't a director in radio, television, or film who hasn't said at one time or another: "Let me hear something that is really different!"

This chapter examines the creation of sounds and the utilization of library effects. Using library effects should not be considered either compromising or unimaginative. As you will see, rarely does an artist have the luxury of time to start from scratch with every sound.

The most challenging aspect of the art of sound effects is in the creation of sounds. It is one thing to put together a "different" sounding car crash, but it is quite another to create the "sound" of sunlight.

As we've learned, sounds may be either natural or characteristic. If a sound is characteristic, it can be broken down further into imitative and interpretive sounds.

Imitative Sounds

One of the basic methods used in creating a sound is to imitate it with another sound that is similar in physical properties. If you want the sound of footsteps walking in the snow, you might use cornstarch (see Figure 6–1). For the sound of a horse's hoofbeats, coconut shell halves work well, as does the end of a plumber's plunger. For the sound of "boiling blood," water lacks the proper consistency. However, Coca Cola syrup works nicely. The thicker the consistency of the syrup, the more closely it suggests blood. Plastic flash cubes are excellent imitative substitutes for ice cubes. Our last example of imitative sounds is somewhat more marginal. What, for instance, would you use for the sound of a human body being turned inside out? The artist who has the luxury of a large budget, a great deal of time, and a very strong stomach may want to go to a slaughter house, tape a butcher skinning an animal, and then spend the necessary time editing these sounds to match picture or dialogue. Not an easy task. For the rest of us, wet thin surgical gloves are imitative enough. By stretching and twisting the wet gloves with excruciating slowness, the imitative sound can be very convincing. I use the words "excruciating slowness" for a purpose. Any sound effect, if done manually, must be done with intense concentration. Simply twisting a rubber glove will convince no one of the agonizing and unbelievable pain that a sound such as this should suggest. (See Figures 6–2 through 6–4.)

Interpretive Sounds

Interpretive sounds are the audile artist's dream. An audile is a person who can look at an object and instead of imagining what function it might serve instead considers its probable auditory quality. To an audile, a piece of cellophane represents the sound of a crackling fire; a cork dipped in kerosene and rubbed on glass is a chattering monkey or squealing rat; buckshot slowly rolled on a bass drum is the sound of surf. Although not all of us have these special talents, by understanding more about the methods of creating sounds, we might take away some of the mystery of creating such interpretive sounds as that of sunlight.

FIGURE 6–1 "SOUND: WALKING IN SNOW." John McCloskey, CBS, New York, fills a large shallow box with cornstarch in preparation for a long trek through a radio winter's snowstorm. Normally, steps in snow were done by rhythmically squeezing a small chamois pouch of cornstarch by hand. Although this made a superior sound, it demanded the use of both hands, and on a very busy show, this presented problems. Therefore, out of necessity, artists working alone chose the above-described method of walking through snow. Interestingly, even today this same method is used on Foleying stages in television and film post production houses. (Photograph courtesy of CBS.)

Acquiring Sounds

The sources of sound effects are as varied as the needs. Whether you are taping your own sounds or using the effects in a library, there are three ways in which to create a sound effect: manually, electronically, and vocally. The method you choose depends on the type of effect, how it is to be used, and how much time and money are available. Does the effect have to be ready "yesterday"; do you have the budget to hire a sound designer? The term *sound designer* is relatively new in films and is somewhat self-proclaimed. It can indicate anything from someone who has designed the sound effects for a film to someone who has overseen all of the film's sounds, including dialogue and the integration of music. To have the

FIGURE 6–2 A classic example of an imitative sound effect. In the movie, *Friday the 13th Part VI, Jason Lives*, the script called for a large bookcase to crash to the floor. However, the director, Tom McLoughlin, insisted on hearing more than that. He wanted the crashing sound to be "wood-splintering." As a result, the sound designer, Dane Davis, had this large bookcase hoisted to the ceiling. Moments later, he and his assistants gave those two ropes a quick tug, and the bookcase came hurtling to the floor. (Photograph courtesy of Dane Davis.)

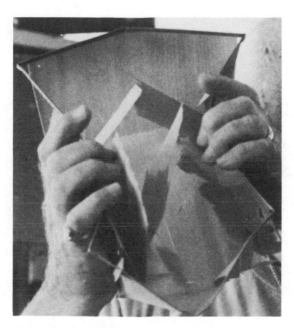

FIGURE 6–3 For that same sound of a bookcase crashing to the floor, the more modestly budgeted television soap would accept the sound of the artist slamming his shoulder heavily into a wooden sound effects house door, with the "wood-splintering" effect provided by a berry basket being crushed close to the mike in sync with the impact of the shoulder against the door. (Photograph courtesy of Ray Erlenborn.)

luxury of the services of a sound designer such as Ben Burtt, who spent almost a year working on the sounds for *Star Wars*, is advantageous to say the least. Admittedly, it is also a rarity to have this much time.

LIVE SOUND EFFECTS

Live sound effects are used when objects have to be physically manipulated to produce a desired sound. Interestingly, the manner in which the effects are created is the same whether they are to be used for radio, television, films, stage plays, commercials, game shows, or home videos. All that is required is a microphone, some technique, and a great deal of imagination.

Often it is possible to extract a number of different sounds from one effect. An ordinary small door buzzer, when held firmly on a rubber pad, will sound exactly as it is expected to sound—like a door buzzer. But take away the insulating qualities of the rubber pad, press it against a large hollow box, and the small buzzer will now have a larger, more strident sound. Simply by holding the buzzer in the air and manipulating the spark

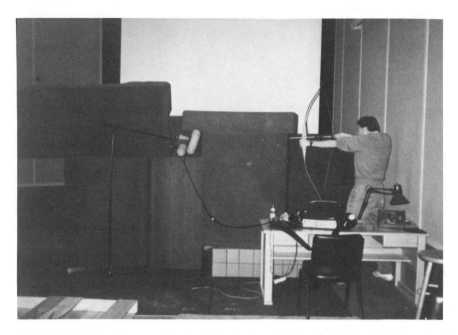

FIGURE 6–4 Another example of an imitative sound. Here, Tom Hammond draws back on a bow in preparation for shooting an arrow into a wooden target. As you can see, the artist has built a corridor of baffles (sound absorbent screens) in hopes of obtaining a good "whooshing" sound for the flight of the arrow and a good "thunk" for the hit. (Photograph courtesy of Dane Davis.)

gap, you can produce a very convincing fly or mosquito sound effect. Clearly, by experimenting with the components of the sound and its acoustical properties, you very often can create such unusual sounds as that needed to satisfy even the requirements of "an ungodly screech" (see Figure 6–5).

Keep in mind that it is not possible to produce the desired sound effect every time you try. It might be necessary to experiment for hours to achieve the proper sound; nevertheless, while experimenting, continue to record your efforts. Once you manage to get just the proper sound, it will be on tape. And if you play it back a thousand times, it will never vary until you edit it in some manner. Of course, *recording* a sound live and *doing* a sound live are two entirely different matters, as you are about to see.

Creating Sounds Live in the Studio

Creating sounds live is very often not only the best way, sometimes it is the only way. However, creating sounds live in the safety of post production, where an effect can be done hundreds of times until it is exactly right,

FIGURE 6–5 Acquiring sound effects necessitates an ongoing awareness of sounds and experimentation. For the film, *Abyss*, the script called for "an ungodly screeching sound." To achieve this effect, Dane Davis, the sound designer, recalled working on another film where the special effects artist placed a large chunk of dry ice, which he had been using for a fog effect, on an old metal table. The molecular reaction between the dry ice and the unpainted iron tabletop produced an ear-splitting screeching sound that Dane never forgot. Knowing that sound would be exactly right for *Abyss*, Dane instructed his assistants, Todd Toon and Blake Leyh (shown here), to recreate this sound. Listening and filing away sounds is an extremely important part of a sound effects artist's job, because waiting to search for a sound only after it has been asked for is usually too late. (Photograph courtesy of Danetracks.)

is not the same as creating this same sound in the television studio, where things are expected to be done right the first time.

Therein lies the tremendous advantage that effects on tape have over effects that are done live. With the taped effects, you will hear the same sound in the exact same manner every time it is played. With live effects, there are never any guarantees. This is especially true of creating sounds that require a certain amount of pressure or friction to produce a desired effect.

In school we have all experienced that unforgettable sound when someone was writing on the blackboard and the chalk suddenly made that terrible screeching noise. Yet if you were to try to repeat that sound on cue, you would find that the majority of your efforts would be unsuccessful. Even when you were successful, could you possibly repeat the sound in the exact same manner time after time? Obviously, a sound as difficult as this to create with any amount of consistency should never be attempted live outside of a post production room. If this sound must be done in the studio, it should be done on tape.

Despite their lack of consistency, very often doing effects live is the only way of doing the effects even under such precarious conditions as live-on-tape in the television studio.

To illustrate how versatile manual effects can be, I have included part of a pantomime sketch I wrote for Dick Van Dyke. The stage was without a set or props, and all communication between the comedian and the audience was done by pantomiming and sound effects.

> HUSBAND:
> ENTERS A LITTLE TIPSY FROM TOO MUCH
> PARTYING

SOUND: CLOCK TOWER STRIKES THE HOUR OF TWO LOUDLY
> HUSBAND:
> GESTURES FOR THE CLOCK TO BE QUIET . . . AS
> HE DOES, A DOG BEGINS BARKING

SOUND: DOG BARKING
> HUSBAND:
> LOOKS OFF AND FRANTICALLY MAKES
> "SHUSHING" GESTURE WITH FINGER TO LIPS

SOUND: CLOCK CHIME AND DOG BARKING GOES OUT AND FADE IN
    CRICKETS CHIRPING SOFTLY
> HUSBAND:
> DOES A SIGH OF RELIEF, TAKES TWO STEPS,
> AND STOPS. TAKES HAND AND TRIES "DOOR
> KNOB" CAUTIOUSLY

SOUND: FOLLOW ABOVE ACTION
> HUSBAND:
> (DOOR IS LOCKED!) PATS CLOTHES LOOKING
> FOR HOUSE KEYS

SOUND: FOLLOW BUSINESS
> HUSBAND:
> FINDS KEYS; SQUINTS TO FIND KEY HOLE,
> BUT IT'S TOO DARK; TAKES OUT MATCHES
> AND STRIKES ONE

SOUND: STRIKING MATCH

> HUSBAND:
> IN STRIKING THE MATCH HE DROPS KEYS,
> FIND THEM, PICKS THEM UP, AND JUST AS
> HE ATTEMPTS TO INSERT KEY IN LOCK,
> MATCH BURNS HIS FINGER AND THE SHOCK
> CAUSES HIM TO MISS THE KEY HOLE AND
> PUSH THE DOOR BELL!

SOUND: LOUD DOORBELL

> HUSBAND:
> PUTS HANDS OVER DOORBELL BUTTON IN AN
> EFFORT TO SILENCE THE SOUND, BUT THE
> DAMAGE IS DONE. LISTENS FEARFULLY, BUT
> HE'S IN LUCK—HIS WIFE DIDN'T WAKE UP!
> TURNS DOORKNOB EVER SO SLOWLY AND
> THE DOOR CREAKS OPEN. STOPS AND LISTENS
> FEARFULLY. INCHES THE DOOR OPEN A
> LITTLE MORE ... AGAIN IT CREAKS. FINALLY
> MAKES HIMSELF AS THIN AS POSSIBLE BY
> SUCKING IN HIS STOMACH AND ENTERING
> SIDEWAYS THROUGH THE DOOR; SLOWLY
> CLOSES IT.

SOUND: FOLLOW ABOVE BUSINESS

> HUSBAND:
> DOES SIGH OF RELIEF THAT WIFE HASN'T
> HEARD HIM. GETS A COCKY SMILE ON FACE.
> THE WORST IS OVER. TAKES BIG CONFIDENT
> STEP ... RIGHT ON THE CAT'S TAIL!

SOUND: CAT SCREECH

Although this is only part of the pantomime sketch, it is enough for you to see why manual effects were so important to its success. The effects of the tower clock chime, the dog barking, the crickets, the doorbell, the cat screeching, and even the match strike can all be on tape because these effects do not need any special treatment. The comedian is merely reacting to them. All the effects that are indicated by "follow the action" or "business" should be done manually in order for the comedian to have the freedom to govern his moves in accordance with the reactions he is getting from the audience. Such things as when to do that next little move with the imaginary door or how much to do should be determined by the comedian's sense of timing and not limited to the type or length of prerecorded sound effects.

Some Disadvantages of Creating Sounds Live

Unlike cart machines, which permit dozens of effects to be played merely by pushing buttons, manual effects require the use of hands or feet. Insamuch as sound effects artists are merely human, he or she can perform just so many effects at a given time.

Consider the artist working on a daytime Soap in a television studio, doing effects in sync with the actress's moves. Keep in mind that these effects are all done in front of an open mike and that every sound, intended or not, is being heard. If, for instance, the actress is opening a small wall safe that is constructed of wood and the director wants the more realistic sound of a metal safe, the artist must sync the metallic sounds with the actress's moves. In order to do this, the actress must make her moves deliberate enough (without being obvious) for the artist to follow the action precisely. Naturally, the more effects that the artist has to sync, the more attention the actress must pay to the order in which the effects are done. For instance, if the artist has to synchronize with the actress the sound of the safe's dial being turned, the artist should know that the actress is going to turn the prop dial four turns to the right, three turns to the left, and one turn to the right. Not that it's important to know whether the actress is going to turn the dial to the right or left—rather, it is important to know how many times. Doing effects live requires anticipation. If the artist waits until the actress makes her move, the effect will usually be a trifle late. Therefore, the actress should make the same moves in the same manner on air as she did in dress rehearsal.

Using a cart to do an effect such as dialing a safe would be extremely difficult. Asking an actress to do the same number of dial turns is one thing, asking her to make the turns the same length each time is quite another. Even if you made a loop of dialing, simply fading the dialing sound in and out would lack the fast attack and decay that can be provided only by a live dialing sound.

A final disadvantage of creating sounds live is the fact that not everyone can do them—certainly not with the same amount of proficiency. Live effects—and at this point I'm not referring to doorbells, telephone bells, or other mechanically operated effects that only require the pushing of a button to operate, but rather the manual effects that require dexterity, imagination, rhythm, and timing—demand a good "touch." It isn't by accident that some of the better sound effects artists were, and are, former musicians or dancers. In radio, there were two types of artists: those who were adept at cueing up records and those who did the live effects. In film, it is the same way. There are the editors who lay in the effects either on tape or film, and there are the Foley artists who do the synchronous work in post production. Eating corn on the cob in such a way as to make it sound like rats gnawing down a door is not a talent that everyone possesses—or, for that matter, wants.

COMPUTERIZED SOUNDS

Using sound generators became popular in the early 1950s. At that time, Don Foster, a sound effects artist, introduced his electronic "Foster Gun" (see Figure 6–6.) In addition to supplying various types of gunshots, ma-

FIGURE 6–6 The "Foster gun." Although today's computerized sound generators are more sophisticated, the manner in which they manipulate an envelope of sound remains basically the same.

chine gunfire, 40-mm antiaircraft fire, cannon fire, explosions, and ricochets, Foster's sound generator was capable of being activated by either manually controlled buttons or sound pressure waves entering a microphone.

This electronic "shot machine" seemed to be the answer to all of the criticism that live television's gunshots were receiving. In those days, an actor firing a blank cartridge that was large enough to make a realistic sound presented a real danger to other actors, because the paper wadding used to hold the powder in the casing would often be discharged with the impact of a fiery pellet. Therefore, most actors refused to use these large-calibered blank pistols in their scenes. To maintain the realistic look of flame and smoke coming from the pistol, a much smaller calibered blank was used, but it produced a very ineffectual sound.

Therefore, it was up to the sound effects artists to sync a realistic sounding gunshot with the actor's movement of pulling the trigger. To match the smoke and flame coming from the pistol took a great deal of anticipation, skill, and luck.

The Foster Gun solved that problem. By tuning the generator to react to the sound of a small blank charge, it would produce a loud, realistic gunshot sound in perfect sync with the smoke and flame emitted by the small, harmless blank. There was only one problem. The electronic gun could be actuated by *any* sound that approximated the loudness of the blank shot. Therefore, the microphone connected to the electronic gun was only turned on immediately prior to the cue, and it was turned off immediately after the cue. Actors were firmly instructed not to talk or make any loud noises when they were about to fire the gun.

On the prestigious CBS program *Studio One*, Artie Strand used this shot machine with great success in a very dramatic scene. As the four shots spewed realistic flame and smoke from the actor's pistol, Artie's shot machine supplied four loud and explosive shots in perfect sync—a triumph for man and machine. As the scene faded out, Betty Furness, the spokeswoman for Westinghouse refrigerators, began her in-studio live commercial.

As Betty extolled the virtues of the refrigerator, Artie was busy listening to accolades from the control room over his headset.

Miss Furness finished the live commercial by assuring the audience of the quietness of the appliance, and to prove it, she gave the camera a sincere, confident smile and said, "You can be sure . . . if it's Westinghouse." With that, Miss Furness closed the refrigerator door, accompanied by an unexpected and yet undeniable gunshot . . . which, thanks to the modern technology of the Foster Gun, was neither a second late nor a beat early, but, to the embarrassment of Westinghouse, in perfect sync with the shutting of Miss Furness' "quiet" refrigerator door.

Synthesizing or Sampling?

Much has been said about the advantages and disadvantages of digital computerized sound. Some complain that it is "too good," that it doesn't have a "real sound," that it lacks "warmth." Perhaps. It is also unequivocally the audio effects device of the future.

In Chapter 3, we learned that timbre, pitch, attack, loudness, sustain, decay, speed, and rhythm affect sound. We also learned that altering even one of these components can be an arduous and time-consuming task. With computerized audio equipment, this is no longer the case.

Most computerized sound equipment has two capabilities—to synthesize and to sample sound. A synthesized sound is basically a sound wave that has been generated by the equipment and then changed in some manner to fit a particular need. This capability is extremely important for futuristic or unknown sounds. Any sound of an electronic nature would fall into this category. By applying the various components of sound to these generated tones, some interesting and unusual sounds can be created.

Sampling sounds is the process of digitally recording a sound and by changing one of its components, attaining a new sound or a series of new sounds.

In the case of synthesizing, sine waves of different frequencies, volumes, and phases are used to construct a sound. In the case of the sampler, the computer reads a series of acoustic properties at a number of discrete points and then combines them to recreate the original sound (sample).

Perhaps the easiest way to illustrate this rather complicated procedure is to compare it with the action of a film or video camera. The camera records a series of images on film or tape that is measurable in frames. When viewed separately, the images are static and simply a recorded moment in time. When played in a projector, each of these individual frames combine to create the illusion of smoothness and movement that is a true representation of the scene as it was originally shot by the camera.

The difference, therefore, between synthesizing a sound or sampling a sound is that a synthesized sound is constructed from sound tones and a sampled sound is that of an actual sound. (See Figure 6–7.)

How They Operate

Since all sound computers are designed basically for the wider musical market, it is not surprising that the means of operation is in the form of a 32-note keyboard. It is the manipulation of these keys that determines if a synthesized or sampled sound is altered or simply replayed from the memory bank.

FIGURE 6–7 A sound effects artist syncing effects with a Synclavier. The effects stored on a floppy disc are displayed on the center monitor. To the left is the tape being sweetened. Simply by programming the effects to match the pictures, the artist is able to sync the Synclavier's SMPTE time code with that of the tape's, so that the effects hit accurately to within 1/30th of a second. (Photograph courtesy of New England Digital.)

There are two methods of storing sounds on these computers—in the main memory hardware or on floppy discs. The main bank normally will store a library of conventional sounds, while the floppy discs will contain sounds of a more specialized nature.

In order for these computers to be of any real service in the post production room, they must be interfaced with the SMPTE time code on videotape or film. With this capability, the computer can be locked to the film or tape for frame-by-frame accuracy, allowing the artist to sync either a single sound such as a gunshot, or do a busy fight scene punch by punch, using just one punch for all the various types of punches. This is accomplished by using the time code and then changing the pitch of the various blows in sync with the action to give the illusion of many different punches.

If, however, you view the scene and feel that there is too much similarity between two of the punches, with some equipment, such as the synclavier, you are able to view that particular wave length on a terminal and edit the sound by a combination of hearing the punch and literally "seeing" the sound. This technique is not unlike deleting letters with a word processor; the difference being that with the sound computer you have the added advantage of listening to the edits.

Suppose you are doing a jungle picture and the director likes one particular parrot sound to the exclusion of all the other bird sounds. By conventional effects methods, making the desired changes would be a nightmare. First you would have to dub off a copy of the entire scene for safety and then edit out the particular bird effect the director requested. Next you would go to the library and hopefully find either that bird as a single or some other bird effect so close in sound that it would be difficult to tell them apart. Even if you were successful in both cases, you would still have to dub and redub the effect to make up the sound of a large group of birds. Of course, what you end up with would be a composite of one bird sound played over and over at different intervals. In order to do the job right, you would have to change the pitch, attack, timbre, sustain, and decay of at least several of the sounds to give your effect true variety. Certainly, no effects artist would go to all that trouble with conventional tape editing methods unless the director insisted on it. Doing so would be extremely time-consuming and therefore expensive.

With a sound computer, the changes could be made in minutes. The desired parrot sound would be fed to the memory, and each desired additional effect would be assigned to a key and its different sound components altered to suit the artist's or director's tastes. Essentially, the computer treats the various parrot effects as a musical composition. Each effect is assigned a "part" that is different in sound from the next effect, but together they form the group of effects that the director requested. It is little wonder that certain film sound editors are being referred to as "sound designers."

Looping

One of the most difficult jobs about making a loop with conventional methods is hearing the splice. This is because sounds rarely remain constant for even the shortest periods of time. Because of this, if you were looping a sound such as an automobile idle, you would allow your effect enough tape so that any noticeable splice noise would be heard only once every minute, and hopefully by then, the dialogue would be so interesting that the audience wouldn't notice. With a sound computer, such as New England Digitals' Synclavier, a looped sound as short as 4 seconds could be played indefinitely without any noticeable splice (see Figure 6–7).

Sequencers

As the name implies, sound computers with sequencer capability enable the artist to repeat sampled sounds in sequential order. If, for instance, you need the effect of footsteps, simply sample several steps, either Foleyed or taped, and you will now be able to make these steps do pretty much as you like. You can slow them down, speed them up, make them heavy or light, even increase them into a crowd. With the SMPTE time code, every step will be in perfect sync.

Some Disadvantages

Digital sound, unlike analog sound, recreates the source signal with amazing clarity. It is also unaffected by generation losses so prevalent with conventional tape. But just as it reproduces the source signal faithfully, any spurious noises will also be faithfully reproduced with sometimes embarrassing clarity.

However, amazing as sound computers are, they are still dependent on what information is supplied to them. Proper communication as to your needs prior to going to post production will save you time and money, just as it will with the more conventional methods of posting effects.

Although these computers have the capability of creating a war out of a few gunshots, it does take time. If you want to treat them as toys, be prepared to pay the price. Whether you work with computers or conventional equipment, whether you work in post production or the studio, nothing contributes more to the final success of your work than being prepared.

VOCALIZED SOUND EFFECTS

It may seem unusual to discuss both the art of doing vocals and sophisticated audio equipment in the same chapter. Yet vocal sound effects in many instances can solve effects problems that are beyond the capabilities of even the most sophisticated computerized sound equipment. Much can be said for the wonders of digital sound, which can transform a single chirping bird into a flock of cheerful songbirds. But can an audio computer give an ordinary horse whinny an English accent?

Who Does Them?

Although *vocal effects* are generally sound effects, they do not have to be done by sound effects artists. The reason for this dates back to the beginning of radio, when it was discovered that not everyone's vocal cords were created equally. Some artists were extremely good at doing dog, pig, cow, and sheep noises, but couldn't do a horse whinny if their job depended on it. And that was just the point. It was decided early on that someone's job shouldn't be in jeopardy just because their vocal structure prohibited them from producing certain sounds.

Additionally, it took more than just vocal capabilities to produce these sounds. It took a self-assuredness to perform under tremendous pressure. It is one thing to be great in the privacy of your shower and quite another to function before a studio audience of hundreds and a listening audience of millions! More than one budding career came to an abrupt end due to "drying up" or "mike fright" when the moment came to perform.

Although vocal sound effects can be performed by either artists or actors, it is generally considered an actor's job function and as a result comes under the jurisdiction of the American Federation of Television and Radio Artists (AFTRA). When you have two pages of happy dog barks and your throat becomes so dry and your vocal cords become so constricted that only raspy choking noises come out, you can appreciate the talents of these "voice people."

Doing Vocals

The requirements for becoming a vocal artist are similar to those for becoming a singer. Both jobs require a highly trained voice capable of producing certain tonal qualities under pressure, and both demand an excellent ear capable of hearing subtle frequency changes. Although the singer has the demanding job of singing the precise notes of a song, the vocal artist's job is similarly taxing.

Whereas it is not unusual to hear someone singing, it is rather odd to hear a person whimpering like a puppy, hissing like a cat, screeching like a hungry baby, squawking raucously like a parrot, or doing any number of other effects that might be required of the artist.

Therefore, one of the first requirements for a vocal artist is the ability to "be" the part they play. If it makes you feel uncomfortable and foolish to snort like a pig in front of a studio audience or a group of your peers, I would strongly suggest another line of work. In order to be successful, you must have talent and be totally uninhibited.

The Genius Of Mel Blanc

Perhaps one of the most talented "voice people" was Mel Blanc. Just a few of his cartoon film credits include Daffy Duck, Porky Pig, and, of course, that loveable, sly old wabbit, Bugs Bunny.

What is amazing is that Mel's career went all the way back to the live days of radio when he was such a valuable asset to *The Jack Benny Show*. To show you the versatility of this amazing multitalented genius, I have included an example of his vocal prowess.

The following is an actual script given to me by George Balzaar, one of Benny's writers for many years. This show aired on March 11, 1945.

SOUND: TRAFFIC NOISES AND AUTO MOTOR

JACK

Ah, there's nothing like an auto ride on a day like this.... Gosh, how times flies ... here it is 1936, and I bought this car in 1924 ... and it was only 10 years old when I bought it.... Yes, sir!

SOUND: AUTO HORN

JACK

I understand the model after this one had the crank in front.... Gosh, what won't they think of next? ... Well, I guess I'll step on the gas and let her out a little.

SOUND + MEL: LOUSY MOTOR UP ... COUGHS AND SPUTTERS ... OF GUNSHOTS ... MORE COUGHING AND SPUTTERING ... MOTOR DIES WITH DUCK CALL

JACK

Hummm ... it's missing a little ... I wonder what's wrong ...

JACK
Well, there's no use sitting here.... Might as
well get out and crank it.

SOUND + MEL: CREAK OF CAR AS JACK GETS OUT ... SOUND OF
CRANKING ... MOTOR STARTS UP ... COUGHS AND SPUTTERS ...
TWO GUNSHOTS ... MORE COUGHING AND SPUTTERING ... MOTOR
DIES WITH DUCK CALL

This mixture of vocal and recorded effects was combined so successfully that the effect for the Benny Maxwell car was one of the most recognizable and famous sounds in all of radio.

Mel's genius at getting laughs from the most improbable sounds did
not go unnoticed by the Benny writers. Therefore, when writing a scene
involving a horse, they decided as a gag to write this cue for Mel.

SOUND: HORSE WHINNY—IT'S AN ENGLISH HORSE, MEL

During all the rehearsals, Mel only did an ordinary sounding whinny,
and the writers were elated that they had finally come up with a cue that
had stumped the genius of Mel Blanc. But like all true performers, Mel
had saved his best for air time. This time when the cue came, Mel regaled
not only the astonished writers, but the studio audience, the millions of
listeners at home, and Jack Benny himself with not just an English
whinny, but a whinny with a Liverpool cockney accent! The studio audience laughed their approval and gave Mel such an energetic burst of applause that it stopped the show. That was the last time the writers wrote
gag cues for the incomparable Mel Blanc.

Personalizing Sounds

Today the application of vocals in television and films is of equal importance. If, for instance, you are doing a cartoon and you need the sound
of a monster roar, you can use any number of combinations of taped animal effects, which can be layered, speeded up, slowed down, played backwards, or otherwise manipulated. However, if you want to *personalize* these
monster sounds so they can be responsive, the quickest way is to use a
vocal effect artist. Whether you use them live or have their vocal effects
sampled and synthesized, their ability to give you the exact effect you want
and to breathe life into not only cartoon characters, but inanimate objects
as well is amazing.

To attempt to accomplish much of what vocal artists do in minutes
can be very often difficult and expensive by more conventional editing
procedures. They are an extremely gifted and creative group of artists and
should not be overlooked as a source of effects, especially if the effects you

are looking for require a personal touch. I cannot help but wonder what an electronic sound library would have provided when called upon to produce the sound of an "English horse whinny."

CREATING A SOUND THAT DOES NOT EXIST

It is one thing to create a sound when you know how it should ultimately sound, but it is quite another to create a sound that has never been heard before. Although what we discussed about the various components of a sound will be extremely helpful once we decide what the sound is, at this point, we have no idea what the sound is going to be or how to go about obtaining it. The sound that we're after is that of *sunlight*.

One memorable sound effect that actually happened in radio involved the *sound of sunlight!* As incredible as that might seem, it was an extremely important cue. The story involved a blind girl who had the ability to hear the most extraordinary sounds. Here is a short scene from that program.

> GIRL
> Oh, Mommy, it must be a beautiful day. I can
> even hear the sunlight coming through the
> window.

> MOTHER
> Hear the sunlight?

> GIRL
> Oh, yes, ... and you can too, Mommy ... please
> try.... Just close your eyes real tight, and move
> close to the window so you can feel the sun, and
> then just listen ... listen very carefully!

SOUND: THE SOUND OF SUNLIGHT STREAMING THROUGH THE WINDOW

If you were the producer of that program, what conventional sound would you have accepted?

When selecting this sound or any highly imaginative sound, your choices are not as limitless as you might first imagine. Simply because you are dealing with such a highly imaginative sound does not mean that you are at liberty to use just any unusual sound. Even though the sunlight to our ears is without sound, the effect we use must convince the audience that this is the sound sunlight would most probably make if the rays indeed suddenly became audible.

Ethereal effects, such as the sound of sunlight, are by their very definition not of this world. It therefore stands to reason that any recognizable effect would spoil this illusion. Additionally, if you use a sound that is identified with one object for another, you are asking the audience to make a comparison. For instance, if you use the sound of an eagle's wings for that of an angel's wings, you have destroyed the celestial qualities of the angel by giving it a sound your audience recognizes as earthly. Therefore, what you are saying with this sound is, an angel and a large bird have a great deal in common. By so doing, you have destroyed the ethereal mystique of the angel.

Onomatopoeia

Onomatopoeia is the formation of words to imitate sounds. Some examples include the following: bow-wow, growl, clap, screech, thud, boing, twang—and the breakfast food favorites—snap, crackle, and pop. All these words imitate sounds. Quite often, this is the only clue you are going to get when dealing with fantasy sounds.

Applying Onomatopoeia to Foreign Languages

Obviously, the foregoing examples are all English language words. Although it is perfectly obvious to our ears that an object entering water creates a *splash,* how is this perceived by a non-English speaking listener?

In Japanese, the word for splash is *dobun* (do-bun'), and in Spanish it is *salpicar* (sal-pi-car'). Further evidence of the differences in the way we "hear" certain sounds is revealed by the words used for the sound of an open hand striking someone across the face. In English we most often say *slap,* while in Japanese the word is *pachin* (pa-chin'), and in Spanish the word is *bofetada* (bo-fe-ta'-da).

Although I have only used three different languages for each of the words, the same disparity in interpretive words would be evident no matter how many languages were sampled.

Having always used the word "splash" to indicate an eruption of water, you may simply have assumed there are no other words to describe this sound. As an experiment, try substituting these new words and you'll discover that with the proper inflections, they mimic the sounds as well as our more familiar words.

Although the words *pachin, bofetada,* and *salpicar* may sound strange to our ears, can you imagine a non-English speaker's reaction to the word *boiiiinnnng?* (See Figure 6–8.)

The Boiiiinnnng Box

An excellent example of the use of onomatopoeia is an electronic gadget constructed by hand for the exclusive purpose of producing only one sound. Although it was used in radio for many years, it did not become

FIGURE 6–8 A "Boing" box.

famous until the sound was featured in a highly successful theatrical cartoon, *Gerald McBoing Boing.*

The construction of the boing box is relatively simple. A guitar "G" string is attached to an electronic pickup head similar to the one on your record player; this in turn is connected to and amplified by a speaker. The pitch of the sound is determined by adjusting the wooden slide in the center of the box. When the slide is closer to the handle, the string in effect is shorter, and therefore the pitch is higher. Conversely, the farther the slide is move away from the handle, the longer the string becomes and the lower the pitch.

The "boiiiinnnng" sound is made by plucking the string when it is at a certain degree of looseness, and then by squeezing the wooden handle at the end of the box, the string becomes extremely taut and the frequency of the sound is suddenly raised in pitch to make a "boiiiinnnng" sound.

Looking For Clues

When dealing with a sound so elusive as sunlight, you must give the sound some sort of logic to help convince the audience. Onomatopeia provides that logic.

In order for this creative process to be effective, you must free yourself of any judgmental thinking and simply jot down any ideas (no matter how ridiculous) that come to mind. The time to be critical is later, when you're evaluating the merits of your creative efforts.

The one word that seems to be most used in association with sunlight is *streaming*. Even the writer of the sound effect cue used that particular word. Although it might be difficult to sell the sound of water as that of sunlight, if you listen closely to the sounds of a gurgling, babbling brook, you might hear something useful.

When you listen to a small stream meandering through the quiet woods, it isn't so much a constant sound of running water that you hear, but rather an occasional soft sound. This is caused by the stream going over and around the various obstacles in its path. At one point it may be a rock or a fallen tree limb causing tiny splashes on the water's surface. These small disturbances are very elusive.

Giving an onomatopoeic name to these brief, watery sounds may be most helpful in our selection of an effect to imitate sunlight's sound. Do these small, momentary splashes really sound like "gurgling" or "babbling," or is there a better, more descriptive sound?

To the artist who did that show, the sound that best described sunlight was a soft "tinkling." After trying and discarding hundreds of manual sounds, the one he finally selected to satisfy the delicate, mystical quality of the blind girl's perception of sunlight was glass wind chimes.

As you can see, the process involved in creating sounds is very demanding. Despite all of the sounds that are in libraries, there will always be challenges—such as creating sounds for sunlight or perhaps an angel's wings.

SUMMARY

1. Sound effects libraries contain both natural sounds and sounds that have been created for specific situations.

2. Imitative sounds are those sounds that have the appearance or consistency of the natural sound.

3. Interpretive sounds are selected for their sound qualities only.

4. Live sounds are those sounds that are manipulated manually.

5. The advantages of live effects are the control and the first generation sound they provide.

6. The disadvantages of live effects are their inconsistency and the amount of hands-on attention they need to produce sounds.

7. Synthesizers are basically sound generators that are capable of imitating other sounds.

8. Samplers digitally record sounds and manipulate the various components of those sounds.

9. Vocal effects can be done live or sampled by a computer.

10. The advantage of vocal effects is their ability to quickly provide sounds of a personalized nature.

11. Onomatopoeia is the formation of words to imitate sounds.

12. The sound effects artist must have the ability to listen to sounds and the imagination to utilize them in the most creative manner possible.

RADIO SOUND EFFECTS

Although dramatic radio is no longer the vital commercial force it once was, it still offers an excellent opportunity to learn new skills and techniques that can be extremely valuable in other media.

A dramatic story performed on radio must be uniquely written, acted, and directed for an audience that is influenced only by what it hears. To do so successfully requires intensive cooperation from the writers, actors, directors, music directors, engineers, and sound effects artists. To better understand this, let us examine more closely the roles of the various arts in radio.

WRITING THE RADIO SOUND EFFECTS SCRIPT

Writing for radio is a unique and exciting experience. Once you have mastered the technique of writing for a listener rather than a viewer, the freedom that radio offers is unequaled in any other medium. Simply telling the audience they are in a remote jungle along the shores of the Amazon River projects them into that jungle. The only ingredients needed are convincing dialogue and realistic sound effects.

Some Rules to Follow

Once you have gotten over the exhilaration of knowing you can set your story in any exotic location throughout the world and even change it a dozen times within the span of one program, there are some rules that must be followed to prevent the audience from becoming hopelessly lost.

Rule 1. Establish the locale of your scene as soon as possible. It isn't enough to write elaborate sound effect instructions for the artist if you don't identify these sounds for the audience. If, for instance you open a scene in our jungle story with river sounds and a boat being paddled, don't wait until two pages later to have the actors mention the word *boat*. This doesn't mean you have to describe the boat down to its color and the manufacturer's name, but the audience must know that there is a boat, who is in the boat, and where the boat is.

Here is an example of how it should be written.

SOUND: ESTABLISH JUNGLE BIRDS, RIVER SOUNDS, BOAT PADDLING, AND FADE UNDER DIALOGUE.

These instructions to the sound effects artists simply mean that the writer wants more sound effect activity at the beginning of the scene to establish where the action is taking place. Then as the actors begin talking, the effects should be faded under the dialogue.

> JIM
> How long have we been paddling, Tom?
>
> TOM
> It seems forever ... my back is killing me.
>
> JIM
> See anything up ahead, Mary?
>
> MARY
> It's hard to tell with all this vegetation growing
> out of the water ... but it looks like we're
> coming to a bend in the river.

As simplistic as this rule seems, you must respect it or your audience will wonder what those noises represent. Remember, they are *listening* to your drama and trying to envision the action, not watching it on videotape or film.

Rule 2. After you've played the scene for several pages, repeat Rule 1. You must assume that new listeners will be tuning in and they too want to know what's going on as soon as possible.

Rule 3. Sound effects are radio's scenery and atmosphere, make them as interesting as possible. Although the writer has complete freedom to select whatever part of the world she feels best suits her story, certain locations require special handling.

SOUND: DOOR OPENS ... PRINTING PRESSES UP FULL

> MARLOWE
> (SHOUTING) Are you Harry Gesner??!!

> HARRY
> (SHOUTING) Who?

> MARLOWE
> Harry Gesner!!!

> HARRY
> (SHOUTING) Maybe I am ... maybe I ain't. ...

> MARLOWE
> (SHOUTING) Too bad you can't make up your
> mind. I just want to ask him one question....

> HARRY
> (SHOUTING) I said I never heard of him!

> MARLOWE
> (SHOUTING) It's worth five hundred bucks to
> Gesner ...

SOUND: PRINTING PRESSES SUDDENLY SHUT OFF

> HARRY
> What's the question?

In this scene, no attempt has been made to fade down the effects under dialogue. Instead, this scene is played against the loud sound effects for humor. A word of caution: Whenever you have dialogue under loud sound effects keep the scene short.

Rule 4. Keep your sound effects instructions simple. Numerous sound effects that don't have the support of narration or dialogue are called *sound patterns*. Sound patterns that call for many sounds that are difficult to identify should be avoided. Many writers like to use sound patterns at the very opening of their show to tease an audience into listening to the rest of the program.

SOUND: CITY TRAFFIC IN BACKGROUND. CAR COMES TO A STOP. CAR
DOOR OPENS AND CLOSES. STEPS ON SIDEWALK UP CEMENT STOOP.
PICKING LOCK. DOOR OPENS AND CLOSES. TRAFFIC OUT. STEPS
UPSTAIRS SLOWLY. DOORKNOB BEING TURNED. DOOR OPENS.
THREE FAST GUNSHOTS. STEPS RUNNING DOWNSTAIRS. DOOR
OPENS. TRAFFIC IN. CAR STARTS AND SQUEALS AWAY FROM CURB.
MUSIC: OPENING THEME

As a teaser, the above sound pattern requires too much concentration from an audience. Furthermore, the sounds of footsteps up a cement stoop (how does an audience know it's a cement stoop), the sound of a lock being picked (that's even more difficult to figure out), are simply too obscure and take too long to identify. In addition, you should never slow the action by doing a "literal" cue such as starting the car. Literal cues are those sounds you should do if you want to be absolutely authentic. To offset these literal cues is something called "dramatic license." Dramatic license allows you to take certain liberties with reality for the sake of dramatic impact. And in the case of the car sound effect, the car starting should be eliminated so that we go right to the car squealing away from the curb.

In order for a sound effects teaser to capture an audience's attention, it must be short, exciting, and to the point.

SOUND: FOOTSTEPS ON SIDEWALK, LIGHT TRAFFIC IN BG
(BACKGROUND), CAR APPROACHING AT HIGH SPEED, STEPS START
RUNNING, CAR SWERVES, BURST OF MACHINE GUN, STEPS
STUMBLE, AND CAR SCREECHES OFF.
MUSIC: OPENING THEME

Rule 5. The main purpose of sound effects in radio is to inform. The original purpose of sound effects in radio was to inform the audience as to where the action was taking place and what the actors were doing without constantly having to say it. This is still the purpose. Unfortunately, too often writers use meaningless sound effects simply to fill time. This was especially true of early radio soap operas.

SOUND: CUP DOWN ON SAUCER

MA
More coffee, Hank?

HANK
Nope, two's my limit, Ma ... unless, of course
you throw in one of those cinnamon donuts of
yours.

MA
I think that can be arranged, Hank....

SOUND: SCRAPE OF CHAIR AND FOOTSTEPS

MA
Got some more warming in the oven.

HANK
Now I don't want you to go to any trouble, Ma.

>   MA
> I'm just flattered you like them.

SOUND: PHONE RING OFF

>   MA
> Now who do you suppose that can be at this
> hour?

SOUND: PHONE RING OFF

>   HANK
> While you're finding out . . .

SOUND: SCRAPE OF CHAIR

>   HANK
> . . . I'll just go help myself.

SOUND: PHONE RING OFF

>   MA
> Coming . . . coming . . .

SOUND: FOOTSTEPS
SOUND: PHONE RING FADING ON

>   MA
> I'm coming, land's sake . . . hold your horses. . . .

SOUND: PHONE RINGING ON. STOPS AS PHONE IS LIFTED FROM CRADLE.

>   MA
> Hello?

SOUND: DIAL TONE

>   MA
> Oh dear, they hung up. All that trouble for
> nothing.

Your can say that again, Ma. This is a good example of sound effects being used when a writer has run out of things to say. All of those phone rings and steps for a party that hung up saved the writer almost a minute of meaningful dialogue.

Avoid Wordiness In Your Cues

If writers do their job properly, the majority of sound effects the script needs and where they're needed will be apparent. Take our jungle river story, for instance.

> TOM
> Think we should be stopping soon, Jim?
>
> JIM
> Between paddling this leaky canoe, swatting
> mosquitoes, and listening to all those dumb
> birds, I haven't had time to think about
> anything!

With descriptive dialogue such as this, all a writer has to indicate is:

SOUND: <u>JUNGLE AND RIVER SOUNDS</u>

Unnecessary sound effects directives serve no other purpose than to take up space.

SOUND: <u>IT'S NIGHT ON THE RIVER, AND WITH IT COME THE MOSQUITOS AND THE INCESSANT CRIES OF THE JUNGLE BIRDS. ABOVE THIS DIN WE HEAR THE RHYTHMIC SOUNDS OF PADDLING AND THE GURGLING SOUND OF THE RIVER.</u>

Poetic directives have no place in a radio script. The listener can't read the instructions written for the sound effects artist, no matter how informative they are. Even if the artist were to follow such descriptions of the scene, if the dialogue ignores the surroundings, the audience will begin to wonder what all the noise is that they're hearing. Furthermore, after reading the dialogue in the script, if a sound effects artist has to have it written down to add mosquitoes, birds, and a "gurgling river," it's time to get another artist.

Don't Hide Sound Effects Cues

During a busy radio show, the last thing the artist wants to do is read a lot of unnecessary words for cues.

SOUND: <u>NIGHT HAS FALLEN ON THE RIVER. THE JUNGLE IS QUIET NOW. ONLY THE GURGLING OF THE RIVER AND THE SOUNDS OF THE PADDLES DIPPING INTO THE WATER CAN BE HEARD. IN THE DISTANCE WE HEAR A RIFLE SHOT. SUDDENLY THE JUNGLE IS ALIVE WITH BIRDS RAUCOUSLY SCREAMING THEIR INDIGNATION.</u>

If a writer insists on writing cues such as this, not only will he be looked upon as an amateur, but his cues will end up looking like this:

SOUND: ~~NIGHT HAS FALLEN ON THE RIVER. THE JUNGLE IS QUIET NOW.~~ ~~ONLY THE GURGLING OF THE RIVER AND THE SOUNDS OF THE~~ ~~PADDLES DIPPING INTO THE WATER CAN BE HEARD. IN THE~~ ~~DISTANCE WE HEAR~~ A RIFLE SHOT. ~~SUDDENLY THE JUNGLE IS~~

~~ALIVE WITH THE RAUCOUS BIRDS SCREAMING THEIR INDIGNATION.~~

Therefore, once effects for a scene have been established, redundant instructions should be eliminated; only new effects should be directed.

ACTING WITH SOUND EFFECTS

Acting with sound effects in radio is, in fact, actually *interacting* with sound effects. Unlike the visual art forms, radio relies a great deal on sounds to inform its audience of what is taking place. The better the actor and sound effects artist work together, the more realistic the scene becomes.

The following fight scene is almost de rigueur with any action story on radio. What you are about to read Is what the writer put on paper.

SOUND: TWO FAST GUNSHOTS

> GOOD GUY
> What are you going to do now that your gun is empty?

> BAD GUY
> How about this?

SOUND: BOTTLE SMASHING

> GOOD GUY
> If you're no better with the bottle than you were with the gun ... you're liable to cut yourself.

> BAD GUY
> The only person I'm cutting ... is you!

SOUND: LUNGES

> GOOD GUY
> You're going to have to move faster than that!

> BAD GUY
> How about this!

SOUND: ANOTHER LUNGE

> GOOD GUY
> Now it's my turn!

SOUND: PUNCHES

> GOOD GUY
> Drop it ... drop it....

> BAD GUY
> Okay ... okay....

SOUND: BOTTLE DROPPING

                    GOOD GUY
        Now that we're even ... one for good measure!

SOUND: PUNCH AND BODY FALL

   With minor variations, this is the way most fight scenes are written. However, radio is a very intimate medium where sound effects artists, actors, and directors work only a few feet apart. In the scene you just read, the actors would keep one eye on their script for actual lines while paying close attention to what the sound effects artist was doing so that they could react appropriately to the sounds. Remember, to the listening audience all the sounds are seemingly done by the person speaking the lines on mike. It therefore makes sense that the actors and sound effects artist work to-gether as closely as possible.

   This is the way that same fight scene would sound after the actors and artist added some of their own touches. Additions are denoted by paren-theses and asterisks.

SOUND: TWO FAST GUNSHOTS ...* (EMPTY GUN CLICKS, ACTOR GRUNTS
     IN DISGUST AND THROWS GUN DOWN TO FLOOR)

*(SOUND: SLOW MENACING STEPS. SCRAPE OF CHAIRS)

                    GOOD GUY
        *(BREATHING HARD) What are you going to
        do now that your gun is empty?

                    BAD GUY
        *(BREATHING HARD) How about this!
        *(GRUNTS)

*(SOUND: KNOCKS OVER CHAIR) BOTTLE SMASHING

                    GOOD GUY
        *(BREATHING HARD) If you're no better with
        that bottle than you were with the gun ...
        you're liable to cut yourself.

*(SOUND: SLOW STEPS)

                    BAD GUY
        *(BREATHING HARD) The only person I'm
        cutting ... is *you!* *(GRUNTS WITH EFFORT)

*SOUND: LUNGES *(QUICK SCUFFLING STEPS ... SMALL TABLE
     OVERTURNS)

                    GOOD GUY
        *(BREATHING HARD) *(SLOW SHUFFLING
        STEPS) You're going to have to move faster
        than that!

BAD GUY
*(BREATHING HARD) *(SLOW SHUFFLING
STEPS) How about this?!

SOUND: ANOTHER LUNGE *(SCUFFLING STEPS; CHAIR KNOCKED OVER)
GOOD GUY
*(BREATHING HARD) Now it's my turn!
*(GRUNTS WITH EFFORT)

SOUND: PUNCHES *(FACE PUNCH ... BAD GUY WINCES IN PAIN AND
FALLS AGAINST TABLE ...)
GOOD GUY
*(WITH EFFORT) *(I've been waiting a long
time for this!)

*(SOUND: FACE PUNCH ... BAD GUY WINCES IN PAIN AND HITS AGAINST
WALL ... MORE STRUGGLE AND GRUNTS)
GOOD GUY
*(WITH EFFORT) Drop it ... drop it ... *(MORE
STRUGGLING AND GRUNTS)

BAD GUY
*(CHOKING) Okay ... okay ...

SOUND: BOTTLE DROPPING
GOOD GUY
*(WITH EFFORT) Now that we're even ... one
for good measure! *(GRUNTS)

SOUND: PUNCH *(WINCE OF PAIN FROM BAD GUY) AND BODY FALL.

As you can see, in radio, the more closely the actor and sound effect artist react off of one another, the more realistic the story will seem to the listener.

In reading the revised script, you will notice that the director added a line for the Good Guy: *"I've been waiting a long time for this!"* The reason for this is very important. In radio, when there is a long period of sound effects action, the audience wants to know what is happening. The best way to do this is to give one of the actors a line to say so that the person doing the punching is identified.

Follow the Leader

In the scene you just read, because of the amount of effects and the degree of difficulty is doing them, the actors took their cues from the artists' fight effects and simply provided the vocal sounds to make the scene more realistic. This is the way an action scene such as this should be done.

However, simply because a scene is busy sound effects wise doesn't mean the actors should slow their pace to accommodate the sound effects artist.

In this next short scene, notice how disastrous it would be for the actor to wait for the sound effects.

SOUND: PHONE RINGING AND PICKUP

                             TOM
(TENSE AND ANXIOUS) Yeah! What??!! Don't do a thing until I get there!!! ... I said don't do a thing!!!!

SOUND: PHONE SLAMS DOWN ... RUNNING STEPS

                             TOM
(EFFORT) I've got to make it in time! ...

SOUND: DOOR OPENS ... SNEAK IN BG TRAFFIC ... DOOR SLAMS. STEPS DOWN CEMENT STAIRS AND RUNNING ON SIDEWALK.

                             TOM
I just hope the car starts....

SOUND: CAR DOOR OPENS AND SLAMS

                             TOM
Start baby!

SOUND: STARTER GRIND

                             TOM
Come on!!!

SOUND: STARTER GRIND

                             TOM
Do it now!!!

SOUND: CAR STARTS ... FAST REVS

                             TOM
We're on our way!!!

SOUND: CAR SCREECHES AWAY

It's pretty evident that the success or failure of this scene depends on pace and urgency. Each of the sound effects cues follows word cues from the actor. In turn, the actor needs the sound effect in order to deliver his next line. Any delay on either the artist's or the actor's part could ruin the excitement of the scene.

Therefore, when the pace of a scene is dependent upon the interplay between sound effects and dialogue, let the cues "overlap." By that I mean, neither the actor nor the sound effects artist wait for specific cues, but rather they play the pace of the scene and allow their cues to extend over each other. If there is one cue that must be done at a specific time, the rule of thumb is: because the actor's voice is more adaptable than the sounds on tape, the actor will usually ad lib around a cue so that it times out properly.

THE RADIO DIRECTOR

Unlike television and films, where much of the work on dramatic shows is done in post production, all of the people involved with a dramatic radio show are present in the studio at the same time. The person responsible for everyone working together as a cohesive unit is the director. That person must make cuts in the script for time purposes; coach actors for performance; audition sound effects cues for acceptability; approve music cues; achieve a proper balance between the actors, music, and sound effects; make certain that a proper pace is maintained; and see to it that a program is property timed so that it finishes when it is supposed to.

In order to do this successfully, the director in radio is the absolute focal point of attention. Although the actors might watch the sound effects artist during a difficult scene involving many effects, everyone's attention keeps going back to the director for instructions as to pace, balance, and timing. But because this is radio and some microphones are always open, the director can't give verbal instruction for fear of being heard by the listening audience. To overcome this obstacle, radio uses a series of hand cues to assist the director in relaying information to the cast and sound effects artist.

Some Radio Hand Signals

Whether you are speaking lines into a microphone, doing an effect into a microphone, feeding (playing) an effect electronically, or playing a music cue, someone has to make a decision as to whether or not all these elements blend together smoothly. If the actors are too close to the mike and the sound effects are too off mike, the blend or balance of the program will be disturbed. It is therefore up to the director to listen to the balance of the show and to cue the cast, music, and sound effects. To do this in the most efficient manner, a series of hand signals is used.

a. To cue an actor to speak or a sound effects artist to start an effect . . . *point the forefinger directly at the person.*

b. To increase the volume . . . *elevate the hand, palm up.*

c. To decrease the volume . . . *lower the hand, palm down.*

d. To indicate there is ample time and to slow down . . . *place hands together and slowly stretch them apart.*

e. To increase the pace . . . *extend the index finger and rapidly turn it clockwise.*

f. To warn people to watch for perhaps a new cue . . . *look at the person and point to your eye.*

g. To cue an actor or sound effects artist to be farther off mike . . . *move the hand away from the face.*

h. To move them closer to the mike . . . *draw the hand closer to the face.*

i. To cut an actor or sound . . . *make a slashing move with the index finger across the throat.*

j. To let everyone know the program is on time and running smoothly . . . *touch the nose with forefinger.*

k. When someone on the studio floor wants to question the director as to how the balance of the program is . . . *touch the ear with the index finger.*

l. To let everyone know that the program is doing fine . . . *make circular sign with forefinger and thumb while extending remaining fingers.*

m. To let everyone know that the theme music is about to start . . . *lay the left hand palm down on the right forefinger to form a ''T.''*

These are some of the most important sign cues used in radio. In order to have a smooth running program, everyone must be familiar with them; and because they are so important, everyone concerned, whether or not they are immediately involved with the program, should keep their eyes on the director.

Balance and Perspective

The actors' moves in television and film are carefully planned by the director for the benefit of the camera(s). If an actor doesn't hit his *mark* (proper place), chances are that particular shot will have to be repeated.

In radio, the director doesn't place marks on the floor. Each actor and sound effects artist is expected to mark his script in regard to the proper perspective of his speeches or effects. Perspective deals with distances and has nothing to do with loudness levels. If you are the focus of attention (talking on mike) in the living room, all other sounds and speeches are relative in distance as to how you hear them. Whoever is the focus of attention is said to be *on mike.* If that person is talking to someone upstairs in the bedroom, the person upstairs is *off mike.* How far off mike depends on how big the director wants the mythical house to be. Simply speaking softer and still being on mike does not give a feeling of distance or perspective. In order to achieve perspective, the actor must be off mike.

When an effect or an actor's line is done close to the microphone, the microphone hears a certain sound. When an effect or speech is done away from the mike, because of the loss of certain frequencies and the number of frequencies that do reach the microphone, the sounds are not simply fainter, they have an entirely different timbre. Therefore, moving a sound away from the microphone gives a sense of perspective, whereas lowering or raising the level of a sound doesn't change the quality or timbre of a sound, only its loudness level.

MIXING A RADIO SHOW

A radio studio might have as many as a dozen microphones and various amounts of direct lines, all of which go into the control room and terminate in some type of audio console. Each of these sound sources is patched (assigned) into a particular volume control (fader or pot) that is connected to a meter that indicates that particular sound source's loudness level.

Prior to rehearsal, the audio engineer does a level check to make sure that each of the different faders is calibrated, thus insuring the microphones all have the same loudness level. For instance, if a live sound effect such as a telephone ring is performed the same distance from each of the dozen microphones, they will all have the same loudness levels. Although it is the job of the audio engineer (mixer) to see that the microphones are technically balanced for broadcast or taping, it is the director's job to see that the actors, music, and sound effects are balanced and mixed in such a way that the program is clear and pleasant to listen to.

If the sound effects artist rings a telephone and the actor answers it and begins talking on mike, the listening audience focuses their attention on that person. If, however, the sound effects artist rings a doorbell that is supposed to be downstairs and a considerable distance away at the same loudness level as the telephone that is on mike, the audience will become confused.

Because neither the actors nor the sound effects artist is ever certain as to how their particular loudness levels or perspective levels are affecting the rest of the action, it is up to the director to monitor and correct any deviations in the program's overall balance. (See Figures 7–1, 7–2, and 7–3.)

Dramatic radio is an exciting and creative medium. Simply because it doesn't enjoy the popularity it once did is no reason to overlook all the tremendous benefits that can be derived from an art form that provides so much pleasure for an almost forgotten faculty we possess—our imagination.

FIGURE 7–1 What was a rather common sight during the "Golden Age of Radio" is a media event today. The above picture was taken (and televised) during a live radio show heard coast to coast on Halloween night in 1985. Much of the material was adapted from Stephen King stories, and the production benefited UNICEF. Pictured on the left is the author; on the right is Tom Buchanan, preparing to do some gunshots. Because the program was done in a large auditorium with poor acoustics, people on stage wore earphones to hear the program better and to receive cues from the director, who unfortunately was positioned out of sight. (Photograph courtesy of Cinda Yank Mott.)

SUMMARY

1. Experience in radio—because of its intensive focus on sound—gives anyone entering film or television an added dimension of creative expression. It was perhaps Orson Welles's experience in radio that resulted in his later critically acclaimed innovative use of sound in films.

2. Although radio frees the writer from all visual restraints, it must never be forgotten that the radio audience can "see" only what it hears.

FIGURE 7–2 Live radio drama has long been absent from the airways. But every once in a great while, for a good charitable cause such as UNICEF, the actors and the sound effects artists turn out to revive this almost forgotten art form. Pictured from left to right: Lynn Redgrave, John Houseman, Jean Kasem, June Lockhart, Casey Kasem, Gary Owens, John Carradine, and John Clark. In the background is the author and in the foreground is the director, Richard Orkin. Interestingly enough, this picture was taken just prior to a live broadcast on Halloween, almost fifty years after John Houseman produced *The War of the Worlds* with Orson Welles.

3. Establishing sounds at the beginning of a scene indicates to the audience where the action is taking place. Once the actors begin speaking, the sounds may be slowly faded down so as not to interfere with the dialogue.

4. Sound effect instructions should be informative and concise.

5. Although SFX help an audience imagine the scene's action, setting and sounds must not be ignored by dialogue.

6. In radio, sound effects are the actor's audible movements. Thus, actors and sound effects artists must work together closely to make a scene convincing.

7. Vocal effects provide an alternative to manual or recorded sound effects. They are extremely adaptable and highly distinctive, and they provide a personalized quality.

FIGURE 7–3   The UNICEF program on the air . . . *live.* From left to right are Danny Cook-sey, Norma Macmillan, Jean Kasem, and Casey Kasem. Note how Danny has to stand on a ladder to reach the microphone. It was because of these inconveniences and the labor laws that women often played the part of children in early radio. In this story, Norma Macmillan played the part of a 9-year-old boy. Some things never change. (Photograph courtesy of Michael Tweed.)

8. The director is the center of attention in radio because in addition to making script changes, critiquing actors' performances, approving sound effects, giving all script cues, and seeing to it that the show gets on and off the air on time, the director is responsible for the balance, perspective, and pace of the show.

9. The economics of radio dramas does not permit extensive tape editing. As a result, once a program has begun, communications from the di-rector are given silently through a series of hand signals.

10. The mixing of a show requires the talents of the audio engineer for technical acceptability and the director for dramatic impact. These two, listening in the control room, determine what a radio program will sound like to the people at home.

11. Unlike the visual media, dramas done on radio require only the listener's imagination. To stimulate that listener's imagination, the director, actors, music director, and sound effects artist must all work together for an integrated harmony of sound rather than for individual recognition.

SOUND EFFECTS FOR TELEVISION

Sound effects in the studio are performed under an entirely different atmosphere than that of post production. The clinical world of post production is that of time codes, while time in the studio depends on recognition and is fought for tenaciously by engineers, production people, stage personnel, actors, and artists. The more recognition your problem receives, the better your particular product will be. Because of this, the studio is often a hectic place.

Film producers have never attempted to add music and effects at the time of the shooting, and prime-time television producers tend to follow their lead. However, daytime television producers, because of limited time and budget allocations, must still do as much of the show in the studio as possible.

If you've ever longed for the "good old days" of live television, working on an hour-long soap opera five days a week comes very close. As in film, the only sound of real consequence on a soap opera stage is the dialogue. The difference is, in films, no real consideration is given to sound effects (they will be added later, in post production). On the soaps, in spite of all the emphasis on pictures and dialogue, as many effects as possible are done in the studio.

Obviously, turning out a 1-hour show in the space of a 12-hour day takes a great deal of cooperation between all concerned. To the sound effects artist, writers, actors, and directors are especially important.

WRITERS

In the early days of television, sound effects were so identified with radio that producers shunned their usage for fear of having their shows identified as "radio with pictures."

Even today the average soap opera writer looks upon sound effects as "door knocks, telephone rings, and oven timers to say when Grandma's cookies are done." This being the case, writers are making no better use of sound effects than did their counterparts who labored over typewriters writing for radio's *Ma Perkins, Portia Faces Life,* or *Young Doctor Malone.*

Television is the medium of pictures . . . expensive pictures. Compared with film, it operates on an extremely different precept and a considerably smaller budget. If it is the writer's job to find new and interesting situations and unusual locales, should not that writer be aware of all available options? A very important option is sound effects. Here are some of the reasons why.

Some Functions of Sound Effects

Union rules specify that "all off-camera (OC) noises shall be the jurisdiction of sound effects." This takes care of all the effects that we don't see. Of course, sound effects artists also supply effects for things we do see, such as a telephone ringing or an oven buzzing to tell us Grandma's cookies are done. In short, sound effects are capable of supplying the noises for anything that makes a noise. The difference is, with sound effects you maintain total control. If, for instance, we do a scene with Grandma vacuuming, the sound is easily supplied by sound effects. If you need a vase filled with water, the faucet sounds are synced with the actor's actions. If Harry is down in the basement using his new power tools while his young wife is trying to impress the garden club, all of Harry's tool sounds are supplied by sound effects. Later, when the garden club is out on the patio, with the help of sound effects, we can even grill some steaks. The scene is conducted as follows: Harry puts the raw steaks on the grill; while off camera, the prop department switches the raw steaks with some that have been precooked. With sizzling sounds provided by the sound effects artist, the scene is convincing. If this scene were tried live, the acrid smoke—not to mention fire laws—would make it extremely difficult, if not impossible.

The technique of substituting an operational prop (this includes steaks) with a prepared prop is called "duping," from "duplicating." The technique is extremely useful in cutting costs, and, as in the case of the cookout, makes possible some scenes that would otherwise be impossible.

These are just a few of the ways sound effects can add another dimension to scripts. As you can see, with a little thought, there can be more to soaps than just door knocks, telephone rings, and oven timers.

Avoid Subjective "Big Print"

Almost as important as what sound effects the writer includes in the script are the words used to describe them. Imprecise directives (the big print in parentheses) should be avoided.

One time back in radio's heyday, a carelessly written sound effect cue and the resultant sound effect became a cause célèbre among network officials and resulted in the sound effect being banished from the air waves.

It all started out innocently enough. The writer created a scene that called for the victim to fall from a building to his death below. Nothing very unusual about that. Ever since radio began, victims had been falling off all sorts of objects and from varying degrees of height because of the following directive:

SOUND: TERRIFYING SCREAM AS HE PLUNGES TO HIS DEATH

Perhaps it was a little wordy, but the writers all subscribed to this cue to indicate that they wanted a scream that faded into silence, allowing listeners to use their own imaginations as to what happened to the person doing the falling.

Well, along came writer X, and he decided to embellish on the actor's traditional scream by adding a sound effect coup de grace.

SOUND: TERRIFYING SCREAM FADES AS BODY HURTLES THROUGH THE
    AIR AND HITS SIDEWALK WITH SICKENING THUD

To the casual eye, there may be precious little difference between the victim PLUNGING TO HIS DEATH and the victim HITTING THE SIDEWALK WITH A SICKENING THUD. But let us take a closer look. PLUNGING TO HIS DEATH is the way most writers wished to do in their victims. They left the *doing* up to the expertise of those people responsible for doing such things.

Now writer X comes along and not only indicates what he wants, but also *how* he wants it and what the reaction should be! Coming up with a convincing "thud" that will simulate a body hitting a sidewalk is difficult enough. Now the writer compounds the problem by writing that it must be "sickening"! Sickening to whom? The sound effect artist? The producer? The director? The actors? The audience?

Effect after effect was tried and discarded because of the subjectiveness of "sickening thud." What was sickening to one wasn't thudlike to another. And on and on it went. Finally, an overripe pumpkin was produced.

When it was dropped from a 20-foot ladder onto a marble slab, everyone agreed that the resultant sound was indeed a "sickening thud."

That night after the show, the CBS switchboard burst into lights with calls from irate listeners complaining that the thud was not only sickening but also revolting and disgusting—and how could the network allow children to hear such horrible noises! From that night on, the only time a pumpkin was seen in the sound effects department was on Halloween.

All this furor was caused by a writer who wrote a cue without giving it enough thought and a producer who regarded the written word as a command from the heavens!

Remember this power of the written word when you write your sound effects cues; keep them brief and direct, and above all, omit all subjective wordage.

The Power of the Big Print

Being nonspecific with your sound effect instructions and objective with your language does not mean making a mystery out of your cues! Such cues as the following serve no other purpose but to confuse the interested parties; they have no place in the professional script.

<u>SOUND: WE HEAR SOMETHING UPSTAIRS.</u>
<u>SOUND: A NOISE IS HEARD OUTSIDE.</u>
<u>SOUND: WE HEAR A SOUND FROM THE OTHER ROOM.</u>

If these were instructions solely for the artist, there may be some excuse, but this is never the case. The involved producers, directors, and actors all have a preconceived idea of what that "something," "noise," or "sound" is going to be. Therefore, when the artist comes up with a sound to try to satisfy these vague directions, it rarely matches up with what the director has in mind—or the producer or the actor.

Many directors treat the writer's directives as if they were chipped in granite. When these instructions in the BIG PRINT are vague, it makes them worry that what they approve for a "something," or a "noise," or a "sound" is not what the writer had in mind. That, of course, is ridiculous; because if the writer had anything important in mind, it should have been indicated in the first place!

To avoid all this confusion and waste of time in the studio, make your cues simple and specific and avoid being subjective. Change ambiguous cues to sound effect instructions that everyone understands.

Remember that sound effect instructions are not meant to be prose. They should impart to others exactly what you want to hear in the simplest form. An "<u>OFF CHAIR SCRAPE</u>" is performed exactly the same as "<u>FROM THE ROOM ABOVE, THE SOUND OF A CHAIR SCRAPE IS HEARD.</u>"

Another economical way of writing cues is:

SOUND: OC CHAIR SCRAPE.

Which simply means "off camera chair scrape." Writing your sound effect cues in this manner makes them clear to everyone who reads them. But notice what happens when you try to inject suspense into your scripts with your sound effect cues:

SOUND: OC EERIE CHAIR SCRAPE

First of all, a chair scrape can't be "eerie" or anything else; it's a chair scrape. However, to the people stranded in a haunted house, the chair scrape may indeed sound eerie, but then so would every other unexpected sound. It is up to the writer to put them in that frame of mind. Once that is done, the writer won't have to rely on subjective sound effect cues to convince the producer how scary the script really is.

Write Support for Your Cues

The next most important aspect of writing sound effect cues is to give them support in the body of the script itself.

Unidentified sounds, especially if present over a long period of time, draw attention to themselves and very quickly become bothersome. Sounds should support and emphasize a scene, not divert and confuse the listener's attention.

Let's examine the following scene:

SOUND: WE HEAR A LIGHT SPRING RAIN FALLING AGAINST THE
       WINDOW. CARRY IT THROUGH THE SCENE UNDER DIALOGUE.

> MARY
> More coffee, Jane?

> JANE
> Just a spot.

> MARY
> Have you heard the news? Helen's pregnant.

> JANE
> Again?

> MARY
> I'm afraid so.

> JANE
> I don't know how that poor woman does it. How
> many does that make?

> MARY
> Four ... five ... I've lost count.

SOUND: MANTEL CLOCK CHIMES

> JANE
> (CHECKING HER WRISTWATCH) Well, she
> should be arriving soon.
>
> MARY
> With *all* her children?
>
> JANE
> (PUTTING DOWN CUP) I certainly hope not.
> Even in good weather when her children *can*
> play outside, they always seem to be under your
> foot . . . but in this wretched weather we've been
> having . . .
>
> MARY
> (LOOKING TOWARD THE WINDOW
> FRETFULLY) Will it ever stop raining?!

Viewers across American heave a huge, collective sigh. So *that* is what that irritating hissing noise was—*rain!*

Let's review the writer's sound effects instruction for the artist:

SOUND: WE HEAR A LIGHT SPRING RAIN FALLING AGAINST THE
WINDOW. CARRY IT THROUGH THE SCENE UNDER THE DIALOGUE.

In addition to being too wordy, the writer has gone to such extremes in describing the rain that in the writer's mind the job is done. She not only knows that it is raining, but what kind of rain it is and even that it is falling against the window. What the writer is forgetting is that sounds come from a source, and unless that source is identified, many sounds are extremely difficult to recognize. Rain is one of them. Unless this sound is given a great deal of support, it can soon become very annoying to the audience. It is for this reason that rain in films, radio, television, or the stage is seldom heard without the readily recognizable sound of thunder.

Background sounds that carry under a scene are for the purpose of establishing some type of mood. They replace the more recognizable but obvious music cue. But because the writer was so specific with the description of rain, that is not the end of her responsibility. Why did she include this sound in her scene in the first place? If she never intended for it to mean anything, she was better off leaving it out. Long descriptive sound cues make pleasant reading, but little else. Unless you want to go back to the silent screen days where the audiences had title cards to read, either acknowledge the sounds you are asking for or leave them for a time when they are important. Television, unlike films, does not have the budget for extensive special effects or location shootings where the weather accommodates the script; in television the script must accommodate the weather—or in our case, the sounds of the weather.

Getting back to our short scene, with just a few changes, see if this doesn't solve our sound effects problem.

SOUND: RAIN

> MARY
> (CLOSING THE WINDOW) Just look at it rain!
> Don't weathermen ever get embarrassed with
> their predictions?

> JANE
> Not the one I watch ... he's too busy telling
> jokes.

> MARY
> (MOVING TO STOVE) More coffee, Jane?

And we're back on script. The difference is—and it's a big difference—the audience knows that the sound in the background is rain. Even the most identifiable sounds (and certainly rain isn't one of them) need all the help they can get.

Remember, sound effects written with taste and purpose can bring support to your script. Indiscriminately written as crutches, sound effects can become a noisy distraction. The excellence of sound effects is usually in direct proportion to how well the script is written. If the characters and situations are believable, it follows that the sound effects will be believable also.

ACTING AND SOUND EFFECTS

All network contracts specify that off-camera sounds are to be produced by the sound effects artist. This is not done to impede the actor's performance; it is merely a fact of union life. You will find that there are unions that turn on the lamps when you flick a dummy wall switch, unions that move the scenery and props, unions that cause the wind to blow . . . or make it rain or snow . . . that assemble your wardrobe, put on your makeup, style your hair, write your scripts, and direct your moves. Then, of course, there is AFTRA and SAG, which monitor your working conditions and see that you receive your residual checks. In short, television is a conglomerate of unions that over the years have learned to live with each other; and that means that sound effects artists do all off-camera sounds, including door knocks.

I specify door knocks because this seems to be the one effect that actors insist on doing. "That is the dumbest rule I've ever heard"; and "If I don't do my own door knocks, how will I be motivated?" are but two common responses from actors upon learning the sound effects artist will provide the door knocks.

I've already answered the first complaint, now about this business of motivation. First of all, if it is absolutely imperative to the flow of the scene that the actor does his own knocking, the artist will invariably allow the actor to knock. This is done on a situation-to-situation basis so that it cannot be construed as a union "precedence." Secondly, all knocks done by sound effects artists are fed to the stage for the actor to hear. This should be motivation enough for a professional actor. If it isn't, a simple solution is for the actor to feign knocking when the sound effect is triggered. Inasmuch as the actor is off camera, the knocking doesn't have to be synced with the sound. In this way the actor will not only get the feel of the action for his motivation, but, hopefully, when the door is opened, the audience won't see him leaning against the doorway with both hands in his pockets.

Playing the Background Sounds

Thus the next big mental lapse that actors have regarding sound effects. Innumerable front doors have opened to the howls of winter winds only to reveal the character standing there as if he or she were an actor in a warm television studio. Even though they have been advised that it is supposedly cold outside, because they don't hear the effects on the stage, the actors neglect to act as if they are cold; thus the scene ends up looking and sounding foolish. Remember, the only cues normally heard on the stage are the effects needed for direct cues. Door knocks, doorbells, telephone bells, over timers, and doctors' call beepers are some of the cues that fall into this category; but background effects are not fed to the stage.

Therefore, if an actor has any doubt as to what sounds are being played during his scene, he should ask the director during dry block. If it is terribly important for him to hear the sounds, they can be fed to the floor during camera block or, at the latest, dress rehearsal. Of course, on feature film sets, sound will rarely ever be fed to the floor. A director may yell out "Bang" for a gunshot, or "Boom" for an explosion, but that will be it. Insisting on hearing all sounds is an indication of an actor's inexperience.

Telegraphing

The next most important thing an actor has to keep in mind when working with sound effects is *telegraphing*. This means nothing more than miming a piece of business so that the sound effect artist knows exactly what is being done. This is extremely important when sounds must be synchronized with action.

The best way to telegraph is to make one's moves a little broader. For instance, if the action calls for starting a car, the actress should lean in slightly and make a minor movement with her right shoulder, as if turning on the ignition. If she is supposed to be pulling her car off a busy highway, she should look in the rear-view mirror and use her turn signals; both of these pieces of business are not only familiar to the audience, but they indicate to the artist that the actress is about to "stop."

When an actor turns on a waterless faucet, the director obviously will be shooting high so the audience cannot see the absence of water. To make the scene more believable, the actor can make a slight movement with his shoulder to indicate he is turning the faucet on or off.

The secret of good communication between actor and artist lies in the actor's awareness of where the camera is shooting. If an actress is to turn on a stereo and the camera has her full-length, head-on, and nothing is blocking her, a simple flick of the finger miming hitting the play button is sufficient. However, if the camera is shooting her from the rear, a slight up-and-down movement with her shoulder is needed.

Consistency of Movement

When operating a dummy prop with which sound effects must be synced, actors should make the same movements in the same place every time. They should not, for instance, make an up-and-down movement on top of a cassette tape recorder to start it in dress rehearsal and then change to a turning motion on the side of the machine when on the air. If a change is necessary, this information should be conveyed to the director, who will in turn inform the sound effects artist.

Slaps, Punches, and Mayhem

Fight scenes are the most difficult actions for a sound effects artist. Back in radio, everyone got punched on the jaw with a resounding smack of a sound effects artist's fist hitting a leather wallet. The bigger the fight, the more artists and wallets. In addition to punches were overturning tables and chairs, breaking bottles, and falling bodies. This was fine because of the number of available artists; besides, the listening audience had no idea who was getting hit where, or for that matter, with what.

Today, with the popularity of violence, head punches aren't enough. One fight might include karate chops, body punches, groin kicks, head slams against various immovable objects, and/or bottles and chairs being broken over various body parts.

These fights are usually staged by stunt people from the film industry who are accustomed to having the sounds laid in later in post production.

This explains why they rarely inform anyone of a change in the order of punches unless it concerns camera angles. However, it is important that the artist know about any such changes. Furthermore, the artist must know what type of punches will occur where so that he can provide the appropriate effect at the correct moment. Head punches, kicks, and body blows all have different sounds. Whereas a head punch that is either early or late can be slipped (moved) in post production work, a body punch that has been substituted by actors/stunt people for a head punch will result in a totally incongruous sound and will require a great deal more time in editing.

The artist might have as many as ten different carts loaded for this fight. Between carefully watching the fight on the monitor and pushing the right buttons for all the different types of punches and crashes, there is no time to react to a punch that starts for the jaw and ends at the solar plexus. Unlike film, where the sounds of the punches are laid in frame by frame over a period of days in post production, in television, the artist has only that brief time in the studio to get it down on tape.

Keeping Movements Natural

A warning to actors: Never let your performance be affected because of a sound effect cue. If you have to telegraph so broadly that the whole world is aware of what you are doing, you are not doing yourself or the show a service. One actress who was so afraid of picking up a ringing telephone for fear that it wouldn't stop ringing would put her hand on the phone and wait for a pause between rings to pick it up. This is totally unnecessary. We are no longer doing live television, and if a mistake occurs in the timing of a cue, it can be slipped in post production. What cannot be done in post production is to make the actors' moves seem natural and normal.

In summary, keep your moves clean and direct, and perform them with purpose. Even though you won't hear the effects on the floor the majority of the time, it's the manner in which you move that will make the sounds convincing; after all, isn't that what acting is all about?

DIRECTORS AND PRODUCERS

Regarding last-minute sound effects, directors want to know how soon they can have them while producers want to know how much they will cost. Both mistakenly assume that every sound effect ever recorded is either at the artists' fingertips or can be made up in a matter of minutes. Neither assumption is true. As inventive and imaginative as most sound effects artists are, they must be given the proper amount of time to prepare for a show. The question is, what is a proper amount of time? To help in this matter, let's examine the factors that cause one script to take more time to prepare than others.

Degree of Difficulty of Sound Effects

Determining the degree of difficulty of a script, and therefore the preparation time needed, can be very hard. The number of effects; how, when, and where they fall; and the availability of the effects all influence the amount of time needed to prepare for a show.

Number of Effects

The number of effects on a show is the least important factor in determining a script's difficulty. You can easily handle ten effects on one page if there is sufficient separation of the effects. Conversely, you might have just five effects in an hour-long script; but if they overlap, they can cause problems. For example, in radio, I saw two effects give a veteran artist fits. One effect was a door knock; the other was the sound of taking dictation on a typewriter. Although done manually, neither effect would normally have been difficult. However, the effects overlapped. Imagine knocking on the door with one hand and simultaneously typing with your other hand! Try it.

How effects are executed depends on how much preparation time you have, what effects are available, and where they occur in the script. If, for instance, you wait until the last minute, you may indeed have no alternative but to use a manual typewriter and physically knock on the door. However, given the time, one or both of these effects could be done on reel-to-reel tape or cart machines, thus saving valuable studio time or even having to do the effects in post production.

How, When, and Where Effects Fall

The least difficult scripts are the ones that contain familiar effects that don't conflict with one another and don't have to be synced with action. Changing any one of these conditions will greatly alter the script's difficulty.

If, for instance, the script calls for your characters to be flying in a light plane during a thunderstorm and they turn on the radio for a projected weather report, this is considered a light script. All that is involved is the plane's engine and the storm effects.

Now let's see how this relatively light show becomes extremely busy with just a "few" simple changes.

First, instead of the actor playing the part of the weather reporter, reading his lines on a separate mike in the studio, the director decides to tape his voice and give it to sound effects to play. (It's a sound effect because the voice comes over the radio.) If it were simply a matter of starting the tape and letting it run, there would be no problem. However, if you want dialogue in between the weather reports, the tape will have to be edited and a leader put between the various lines of dialogue. Additionally, if you want to add static (on cue), thunder (on cue), and a sputtering plane motor (on cue), you will turn what was once a fairly light show into a very busy show—and not just because of the *amount* of effects involved. Unfamiliar effects (the prop plane stalling); overlapping effects (storm, plane, static, stalling engine, voice); and business cues (turning the radio on and off) combined with cues from the director (static, thunder, stalling engine, voice) add up to a very different script than the artist originally read and prepared for.

All of these changes take time . . . valuable rehearsal time. Had the director informed the artist as soon as these changes were contemplated, the artist could have gone to the library and selected precisely the right plane effects, including not just the stalling effect, but perhaps also some short dives and zooms for added drama. By knowing the voice would be cued, the artist could have put the lines on separate carts, freeing himself from having to start and stop the voice on reel-to-reel tape. Furthermore, given the time to put all the cues on a cart machine, the artist could have been freed of having to operate several pieces of equipment at the same time. Finally, if the artist had known of the changes beforehand, she would have had the sound effects available for all the rehearsals.

Even if this preparation time involves overtime costs for the artist, money is saved in the studio. Two or three hours of overtime for the artist outside the studio can hardly be compared with the cost of wasted studio time involving a cast, camera and audio crew, stage crew, makeup and wardrobe people, and production staff!

As you can see, although we have only changed the amount of effects slightly, the *manner* in which we do them has a tremendous influence on the degree of difficulty and the time involved.

Again, the amount of time depends on the complexity of the script and the availability of the effects. While some producers and directors feel that the only effects that can be done in the studio are doorbells, door knocks, phone rings, and thunder claps, others apparently believe that studio sound effects artists have everything at their fingertips, and they have no compunction about requesting at the last minute "a Gold Coast African circumcision rite with and without drums"!

Availability of the Sound Effects

Because of limited network studio space, various networks soaps are often done from remote locations. *Days of Our Lives* is done in Hollywood, while the NBC sound effects department is located many miles away in Burbank.

At the studios, there is an accumulation of literally millions of different sounds—everything from electric transcriptions and 78 rpm records from radio to sound effects edited from EJ (Electronic Journalism) camera tapes. There are also sound libraries purchased from outside companies; in fact, some sound effect companies sell effects by the running foot of tape. If that isn't enough, most departments have excellent tape recorders that are available to record any sounds that are physically possible to produce. If they can't get the "real" sounds, they can be duplicated by mixing other sounds together, slowing them down, speeding them up, making them shorter or longer, or changing their equalization. If all else fails, there are the thousands of manual effects that can be recorded and mixed with other sounds and slowed down or speeded up, etc.

Of course, what normally happens is, once in the studio, as the scene comes up, the producer/director calls over the headset, "Sound effects, we made a change. Mary is going to turn on the television set, and we have a clip (a short piece of tape or film) of a silent film of a Laurel and Hardy movie where an automobile gets sawed in half and then blows up and a whole bunch of junk falls all over the place . . . can I hear that?"

In an effort to be cooperative, the sound effect artist responds, "I don't have anything exactly like that on hand, but if you give me some time . . .," to which, the producer/director mumbles, "Why is it you can never get what you ask for without it being a big deal!"

Such unpleasant episodes can be avoided by the director/producer in several ways. Ideally, and this includes any film inserts *or* post production sweetenings, the artist should view the film *prior* to the taping day or sweetening session. Not only will she see things that need effects (other than the obvious), but she will be given ample time to either get the effects or make up the effects needed.

If viewing the film is impossible prior to the taping date, providing the artist with a detailed account of what is needed is the second best solution. A competent artist will bring what is requested, plus additional effects that might be of use.

Simulating Location Scenes In the Studio

Although shooting scenes at remote locations gives any studio program a more realistic look, the tremendous expense involved makes this practice impossible on a limited budget. Therefore, the next best thing is to stimulate location scenes in the studio.

If, for instance, your story line takes your characters to Venice, Italy, it is perfectly within reason to want to do a canal scene in a gondola. However, you must understand and respect the limitations of television. Whether you use only a studio set, or a studio set plus an actual Venetian canal scene shown on a rear screen, it will very often be the audio that will give the scene's believability its acid test. The mixture of dialogue spoken in the studio and ambient sound effects mixed in the audio booth is of paramount importance.

When the sound effects artist mixes effects with other production sounds (dialogue and music), this is only a reference studio mix for the artist. It has nothing to do with what will eventually be heard on the air tape, which is the responsibility of the audio mixer. On many shows (soaps in particular), the final mix is compiled in post production.

To achieve the final mix, two video machines are used to record the show. On each 1-inch tape there are two audio tracks. On machine "A" are the program dialogue and music tracks. On machine "B" are the program dialogue and the effects tracks. Each is a discrete (independent) track, the level of which can be raised or lowered to effect a good balance. However, it takes more than just a technically acceptable balance to make a scene believable. If the actors don't "play" (be aware and react) to the background sounds in the studio, simply raising their voice levels in post production will make the dialogue louder, but will lack believability. The only way to remedy this problem is to have the actors come to post production and redo their audio tracks by a process called "looping." This entails erasing the dialogue done in the studio and lip-syncing new dialogue to the pictures on tape. A slow, arduous, nail-biting experience.

The human voice is a most remarkable instrument. It is capable of giving single words many different meanings simply by changing inflection, intensity, and loudness. Take the simple word *stop*. When spoken by a harried mother of four young children in a crowded movie theater, it is not spoken loudly, but very intensely. At a crowded beach, the mother would speak louder in order for her children, playing some distance away in the water, to hear her. Now if we were to substitute the voice the mother used in the theater with the voice she used at the beach, it would seem totally inappropriate, despite how we treated the beach sound effects.

This illustration is unlikely because people naturally adapt their voices to their environment, taking whatever action is necessary to be heard in the manner in which we want to be heard. When speaking in a quiet studio, despite what sound effects are being added, actors invariably speak in quiet tones. Therefore, whenever a scene is being played to simulate a location site, the actors have to be told (and reminded) exactly what sound effects they are playing against.

Feeding Effects to the Stage

A simple solution would seem to be to feed whatever sounds you are using to the floor so the actors can react to them. This is unwise for two reasons. When sounds are fed to the floor, they are picked up by the microphones used for the actor's dialogue. Later, in post production, if this dialogue needs to be edited, whatever sound effects that were heard under that dialogue will also be edited. The second reason has to do with the quality of the sound effects as heard over the speakers on the stage. These speakers were selected so that the director could communicate from the control room with the people on the stage; these are essentially public address systems, not appropriate systems for laying sound effects on tape. Although it is true that certain live sounds such as door knocks, doorbells, and phone bells are fed to the floor for the convenience of cueing the actors, the quality of these sounds, for the relative short duration that they are heard, does not adversely affect the production.

It is the responsibility of the actors to imagine the ambient sounds and to play against them. In more violent scenes, such as fires, it is perfectly within reason to have the cast hear the sounds during rehearsal so they are acquainted with what noise levels they are working with. But once the tape rolls, the actors must rely on their imaginations.

Voice levels and energy are extremely important when simulating exteriors, because despite all ingenious editing done in post production, the believability of a scene is still up to the actor.

Know Your Limitations

Injecting a note of realism into a scene that has already been shot is probably the greatest challenge during post production. If sufficient forethought is not given in the studio, what you do in post production comes across as contrived.

If, for instance, you are shooting a scene that is supposed to take place in a stable; it doesn't matter how good the set looks, someone (director or writer) must address themselves to the question of horses. If having them is too expensive or inconvenient, that is fine, but the illusion of the horse being there—just out of our sight—cannot be accomplished convincingly by sound effects alone.

By adding off-camera horse whinnies, you endanger the scene's credibility in several ways. The first involves the actors. As discussed previously, sound effects are not fed to the floor because of the editing problems they might present: sounds that overlap dialogue cannot be edited out. Therefore, unless the actors are cued by the stage manager or somehow know precisely where the effects come in dialogue, these unusual sounds will go unacknowledged by the actors—a poor practice. Furthermore, not only will the sounds upstage the actors by competing with them, they will also distract the audience, who will want to see the source of these interesting sounds.

One solution lies in a simple writing technique called "not opening a can of worms." All that means is, if you don't draw attention to a problem, chances are the audience won't notice it. In the case of the stable scene, if you are not going to show horses on camera, don't open a can of worms by teasing the audience with off-camera horse sounds.

When writers and directors face the limitations of a studio, they make significant contributions to the realism of a scene.

The Problem With Crowds

Filmmakers long ago recognized the problem of recording crowds; as a result, there is a group of actors who make a handsome living doing something called "walla walla." The name derives from the nonsense word "walla," which extras used to say with varying degrees of gaiety to furnish restaurant scenes with an ambient crowd sound without anything specific being said. Today in films this is no longer the case. The atmosphere people still move their mouths and act in an appropriate fashion, but most often their words are supplied in post production by professional "wallawers," who, incidentally, no longer simply say "walla walla," but engage in conversations (even using dialects) that are appropriate to their surroundings.

This film technique is impractical for most television shows because of the time and expense involved. On Soaps for instance, the generic term for crowds is *hubbub.* But because most atmosphere people (formerly called *extras*) who are used in scenes such as restaurants and weddings are inexperienced, it was found to be more expedient for the actors to panto-mime talking and have the crowd sounds come from sound effects. This not only affords the director control of the background hubbub level, it also averts any union problems with the actors. Atmosphere people may talk, but it can't be anything of a specific nature (written lines, for instance). If they do speak rehearsed lines on cue, they are placed in a differ-ent and higher pay scale category called *under five actors* (U5's), where as the name suggests, they may speak up to, but not more than, five lines.

Most Soaps have a library of different crowd sounds that satisfy most situations. If the director is not satisfied with what is available, he or she usually wild tracks a minute or so of nonspecific hubbub from the atmo-sphere people for the sound effects artist to edit into a loop and play under the scene.

ASSISTANT DIRECTORS

On many network studio programs, it is the responsibility of the assistant director to cue sound effects. What follows is some advice for these assistant directors.

Remember that there can be only one voice of authority in the control room. To avoid confusion and possible problems, be certain that all of your instructions and cues are consistent with those of the director.

After you have established what sound effects you want in your scene, the next most important thing is to clarify how you will cue the sound effects artist. There are three basic cues. Although there is no particular order of importance, we'll start with the "sight" or "business" cue. To illustrate this, let's take our old friend, the tea kettle. When someone in the cast takes the kettle over to the sink to fill it with water, the action of turning on the faucet is called business. If the sink isn't practical (not rigged by special effects to produce water), the sound of the water going into the kettle is produced by sound effects. The cue to make this sound comes when the actor does the business of turning on the faucet, and because of this, it is called a *business cue*. Because the action is visible, it might also be called a *sight cue*.

Business Cues

The sound effects artist should be told beforehand what action the actor will perform immediately prior to the sound effect. If, for instance, the action calls for a slap, the artist should know which hand is going to be used and whether it is going to happen on a look, a camera angle, or a word cue.

The only time a business or sight cue should come from the control room is when the artist can't see the action on the sound effects monitor. Remember, many sound effects rooms have only one monitor—the line or air monitor. Therefore, if the action isn't being seen on that monitor, it isn't being seen in the sound effects room. However, well-equipped sound effects rooms have two monitors—the air monitor and the preset monitor. The preset monitor shows what camera is going to be taken next. Between the two, the action is pretty well covered. But keep in mind, the assistant director sees all the camera monitors, plus the preset and the program. So if you see that the action isn't being seen on the monitors available to sound effects, by all means cue the sounds. Of course, if you really want to draw attention to your inexperience and in the process make yourself extremely unpopular with the sound effects artist, you can make a practice of cueing all business or sight cues that are seen equally as well in the sound effects room!

Word Cues

Similarly, unless there is a last-minute change, there is no need to cue a *word cue* from the control room, for the sound effects artist has a script.

MARY
I can't understand what . . .

SOUND: PHONE RING

These particular cues can be pretty tricky, especially if the control room is sound effects cueing-happy. One of the problems is very evident above. The line is so short that by the time the director cues it and the artist reacts, an awkward silence has occurred.

Thus, business cues and word cues should not come from the control room. In both cases, the time delay in relaying the cue is enough to spoil the timing of the cue. Even though the effects can be slipped in editing, there really is no need for it if the artist is allowed to take his cues from the action or word rather than from the control room. Some performers are not as good as others at ad libbing, and if the sound effects artist is late on a word cue, it can leave an embarrassing hole in the scene.

"I can't understand what" is not a long line, and if people in the control room, in their lust for power, insist on cueing the artist, it demonstrates their lack of knowledge regarding sound effects and an actor's timing.

A word cue, especially a word interrupt cue, must by its very nature be anticipated. In the case of our illustration, the phone should be rung after the word *understand*. Taking into consideration the fraction of a beat it takes for the sound effects artist to push the telephone bell, the ring should actually occur after "what."

However, if people in the control room insist on cueing, an unnecessary element has been added that can only slow the cue and force the actor to continue the speech with ad libs, which he or she may not be capable of doing.

It should be pointed out to writers that these interrupted sound effect cues should include *several* throwaway words. An interrupted speech should never end with a word that is important to the story line, for it may be buried under a phone bell.

Pacing and urgency of the actor's lines are important considerations when cueing on the word. The best way to stop the dramatic flow of an excited, agitated, angry, screaming actor is to be late with the word cue. Likewise, being late (or early) with word cues, thus spoiling a comic's timing, will usually result in your hasty replacement.

Marking a Script

When the artist is taking her cues from the script, the person doing the cueing is responsible for relaying relevant script changes; this does not include word changes on pages that do not affect sound effects. However, if whole pages are deleted or if large cuts are made within three pages of a cue, the sound effects artist should be informed.

If several lines of dialogue are to be cut, do not obliterate them with scribbles; simply indicate the cuts by boxing them off. Therefore, if the cut is restored, it is a simple matter of noting the word *good* in the margin of a block cut. Had the cut been scribbled out, a new page (not always available) or rewriting (a messy job at best) would be necessary. A trick I have always found useful is to draw an arrow from the last word of the script that is good to the first word after the cut.

Exterior Sounds

When marking a script to indicate the continuation of a sound or music, a simple wavey line drawn in the margin will show how long you want the effects to stay in. When the effects or music are to go out, you draw two short straight lines.

Timing Cues

The third type of cue must come from the control room. It involves the director's timing, which no one else can estimate. There is no anticipating or delaying the cue; it happens on the director's word. Usually, it involves establishing a mood, and only the director can tell when that happens to his satisfaction.

Although the three cues are quite dissimilar, they all have one thing in common: all should be readied. This should be done in the simplest manner, without excessive and useless information. Remember that in addition to cueing instructions, the artist is listening to the program. This is particularly difficult when the mike is open and the speakers are muted; the artist must wear a headset, with the director's voice in her left ear and the program sound in her right ear. Minimizing instructions therefore makes listening easier.

Readying Cues

A standby ready cue should be given approximately three speeches ahead of the cue. It should be simple and direct, for instance, "Ready the phone ring."

If the cue is to happen at the top of a scene, it is said to be "on the line." This means that after the countdown, the artist takes a beat and either rings the phone on his own or takes a cue from the control room. If the cue involves background sounds or music, "on the line" means the effects are playing when we come out of black.

The important thing for the person who is cueing sound effects to remember is to be informative regarding changes and to keep all extraneous talk to a minimum. There is nothing more annoying to the artist than being innudated with useless chatter that merely affirms the nervousness and inexperience of the person doing the cueing.

The job of cueing sound effects is a little out of the ordinary, especially for the person recently promoted from production assistant. Where else can a person attain the godlike power of asking for thunder and rain and getting it! Of course, the real purpose of the job is to relay information in the most concise manner possible. To avoid confusion, there can be only one person in charge in the control room; all others must limit their authority to cueing and repeating information. If others have creative suggestions regarding sound effects, they should discuss them with whoever is in charge and get that person's approval before making any changes. They must not, I repeat, must not take it upon themselves to make changes.

> ASSISTANT DIRECTOR
> Sound effects, this house is so big and beautiful
> we ought to have a doorbell ... you know, one
> of those things that chimes ... instead of a door
> knock.

> ARTIST
> I'll make the change ... thank you.

SOUND: DOOR CHIME

> DIRECTOR
> Who in blazes changed the door knock to a door
> chime?!

From that moment on, the assistant director has lost some credibility; hopefully, he has also learned a lesson.

Good communication between directors, producers, writers, and all production personnel that have anything to do with sound effects is vital to smooth operation in the studio. Furthermore, it can eliminate the unnecessary costs of posting effects that should have been done at the time of the studio taping. Needs and problems must be made known to the sound effects artist prior to studio or post production tapings. In short . . . communicate.

CONTROL ROOM SOUND SYSTEMS _____

Although the program sound that is heard in the control room during the taping of a show is not the same sound that will be heard after the tape has been edited and sweetened in post production, it is the only sound available. Therefore, opinions are formed regarding what is heard as opposed to what will be heard.

To help you better understand this seemingly double standard of critiquing sound, let us examine more closely the sound systems found in today's television control room.

Speakers

Although meters are the only nonsubjective means of determining a program's audio level, you will nevertheless be influenced by what you hear coming from the speakers. Be forewarned that the sound level you hear can be very deceiving. Because people tend to trust their ears rather than determine sound levels using a meter, when you ask a sound or audio mixer for more program level, be certain they don't simply raise the monitor listening level and leave the level of the actual program unchanged.

Most sound effects consoles are equipped with a monitor switching system, which allows you to hear the program as well as to monitor the levels of the various pieces of equipment in the room. If, for instance, you want to make some critical edits in a piece of music tape, you would probably crank up the level of the monitor so that the edits would be more apparent. When you punch out of the monitoring system and go back to the program feed, if you don't restore the monitor speaker level to its normal position and you don't watch your meter, the abnormal loudness of the speaker tends to drop the level of your sounds. Another problem with excessive monitor speaker level is that you will hear equipment and tape noise, which normally would not be heard on the air or tape.

In addition to producing noise, *any* significant change in speaker loudness level alters the ear's frequency response. Although occasional peaks do not contribute significantly to the listener's sense of loudness, the ear has a tendency to average out these peaks. It should be noted that volume unit meters approximate the listener's subjective hearing and that the levels that are laid down on tape are always somewhat higher.

Furthermore, the speaker's level affects your ear's ability to distinguish subtle frequency changes. For instance, riding a speaker extremely loud when recording music or sound effects will cause a loss in bass when the tape is played back at normal levels, whereas recording with the speaker set at a low listening level causes the ear to be less sensitive to high frequencies.

Program Monitors

All monitor speakers operate on the same principle. They are transducers that convert electrical energy from a source into acoustical energy. When deciding which monitor is "best," four factors must be taken into consideration: (1) quality of the speaker; (2) acoustical property of the room; (3) position of the speaker; and (4) position of the listener.

Quality of the Speakers

Although it is important to have a monitor speaker that will efficiently respond to the input audio signals, slavish devotion to speaker excellency is unnecessary. Back in radio, they had an expression that was commonly heard in New York: "They'll never hear it in Canarsie!" (a city located in the nearby borough of Brooklyn). It was aimed at the directors who spent endless energy and time listening to every insignificant sound that had nothing to do with the story and had absolutely no chance of being heard over the inferior radio speakers only a few miles away in Canarsie.

This is not to say that you should settle for inferior sound. Just remember that the studio monitor speakers are expensive, state-of-the-art equipment with a full dynamic range capable of reproducing frequencies that are quite beyond the capabilities of the average home television set. It is for this reason that most record producers and post production rooms have an additional small speaker that closely simulates the average home receiver, which provides them with a more realistic assessment of what their final mixdown will sound like to the buying public.

Acoustical Quality of the Room

Unless a great deal of thought is given to the design of a control room, it matters very little what quality your speaker represents in terms of faithful reproduction of a particular sound.

A control room may be acoustically perfect in design but suffer dramatically when subjected to the cold glare of reality—one aspect of which is the human factor. For instance, maintaining proper speaker function might depend on limiting the control room to operating personnel. However, try telling that to the producer who has a number one in prime time hit on her hands. Suddenly the number of people in the control room grows from the recommended ten to include extremely important network executives, agency people, guests of the extremely important network executives, guests of the agency people, friends of the producer and relatives of the producer—all of whom breathe, wear absorbent clothing, talk, laugh, and generally disturb the engineer's scientifically designed, acoustically perfect control room. Of course, an acoustically perfect audio balance obviously is not everything in a control room, for I have never met anyone who has become successful by telling network and agency executives or friends of a producer to shut up.

Position of the Speaker

Although I have seen monitor speakers hanging in whatever space there is available, certain locations are superior to others regarding loudness levels. As a rule, the optimum loudness level is reached by placing the speakers at the intersection of the walls and floor, or of the walls and ceiling. Part of the increase in level is due to the loudness increase in the bass and part to the concentration of surface areas from which the sound waves can radiate.

Position of the Listener and Other Distractions

It should be obvious that not everyone in a television control room, because of their particular interests, will hear the same program in exactly the same manner. The writer is annoyed that they cut the best lines from a scene she worked so hard on; and she isn't thrilled with the way the actors are delivering the remaining lines. The set designer wishes the director would have shot the scene from a different angle. With all those actors in the way, no one will ever see his beautiful new set! The music coordinator is wondering why the sound effects are drowning out his music cues; and the jealous lead actor has dropped in to see how many close-ups his co-star is getting. And on it goes.

Despite all the others, there are people in the room who are expected to make audio level critiques.

Because of the noise and distractions in the control room and the positioning of the listeners, most control room speakers are set excessively loud, thus affecting the sound's frequency response. How then can the producer or director make intelligent audio assessments if what they hear in the control room is different than what is heard in the sound effects room or the audio booth? The fact of the matter is, they cannot. Even by clearing the control room and sitting in front of the speaker, with the level identical with that of sound effects and the mixer, they will still experience critical, subjective audio discrepancies. This is normal, and no amount of technical advances will ever change it.

There are, however, a few steps that will make the job easier. First, select an audio mixer and sound effects artist whose credits and opinions you respect and trust, because, above all, good communication must exist. If the show is difficult or unusual, those involved should exchange input. If necessary, effects can be auditioned. Despite all of the state-of-the-art equipment available in today's studio sound effects room, the final decision as to content, level, and perspective is still a subjective one and must be made by either the director or the producer. Any decisions that can be made prior to entering either the studio or the post production room mean money in the bank (and an extended professional career).

SUMMARY

1. Although the sound effects done in the studio are almost always sweetened in post production, the time spent there can be reduced if the director/producer communicates any special needs to the artist before beginning rehearsal.

2. Whenever possible, audition what you consider to be difficult sound effects. Thus you will have what you want during rehearsal and you will reduce your editing time.

3. Although sound effects in the studio can duplicate anything done in post production, be aware that everything takes time.

4. The costs involved in allowing an artist to properly prepare for a show are miniscule compared with those incurred due to last-minute requests.

5. Writers should avoid being subjective with their sound effects cues.

6. Writers, in their search for new and interesting settings, should give thought to not only how a set might look, but also to how it might sound.

7. When writing sound effects into a script, identify a reason for their being there so they aren't merely noisy distractions.

8. Although many sound effects aren't fed to the stage because of editing problems, actors should be aware that the effects will be added and that they must "react" to them.

9. When working with props that are cues for sound effects, actors should keep their moves clean and direct so that the artist can follow them.

10. Directors should be aware that it is never the number of effects in a scene that makes it difficult, but rather, when the effects occur and what type of cues are used.

11. Business cues require the artist to follow an actor's actions by watching the monitor. If the effects involved are manual, the microphone must remain open. With cues such as these, even a few effects can make for a very busy show.

12. Prior to the taping day of a busy sound effects show, the director should schedule a conference with the artist to make certain all the necessary effects will be available. This will result in a better taping, and it will save hours of post production time.

13. Because of editing problems, only those effects needed by the actors for important cues are fed to the floor.

14. When cueing sound effects during a show, keep all talk that does not involve the actual cues to a minimum.

15. Despite the grueling schedule for turning out a show in the studio, with proper communication, the amount of time spent in post production "fixing" effects will be minimal.

SOUND EFFECTS FOR FILMS

Everyone fully understands the need for good acoustics in film, but not everyone knows how to treat sound effects. Although the subject of how, when, where, and even *if* to use sound effects has been discussed for other media, many people feel that sound effects in films are somehow *different*. The only true difference is in the technology. The primary reason for using effects—to inform or to provoke a mood—remains constant.

Although many of the techniques in film sound today have been adapted from television, there are still many differences that merit discussion.

THE PRODUCTION SOUND CREW

Although "sound" and "audio" are synonymous, film gives its sound technicians such titles as "sound mixers" and "sound editors" while television refers to them as "audio mixers" or "audio tape editors." This, I might add, is the very least of the differences.

Film is the world of post production. Although much of television is also done in post production, it is the amount of time and the attention to details given there that separates the two. To understand this more fully, it will be of help to examine some of the job functions that affect sound effects.

The Production Sound Mixer

The first person to record a film's sound once shooting gets underway is the production sound mixer. It is that person's function to record not only the film's dialogue, but also wild tracks and room tone.

Wild tracks. The name given to sounds that are not synchronous with the picture is *wild tracks*. They can be in the form of dialogue, vocal sounds, specific sound effects, and nonspecific sound effects.

If, for instance, the director is shooting a scene involving a motorcycle gang terrorizing some teenagers at a small remote diner, there may be some sounds at this exterior location that are so distinctively interesting that he wants to make a separate tape of them for editing. Perhaps its the squeaking sound of the diner's rusty sign swinging in a breeze. Or maybe it's the loud revving sounds from a particular motorcycle. It might even be the horrified screams of one of the actresses.

Any recording done specifically for the isolation of a particular sound can be considered wild tracking. This does not mean that these wild tracked sounds won't eventually be matched with a picture, it simply means the director wants complete control over the sounds' length and loudness during editing.

Sounds that overlap one another cannot be separated. If, for instance, the action of a scene is moved closer to the diner and attention isn't paid to the sign, the squeaking sounds from the sign may be heard during a scene involving dialogue. If that is the case, and the squeaking sound is of sufficient loudness, it can never be edited out from the dialogue. If the sign isn't in the shot, the mixer must ask the prop person to remove it. If the sign is in the shot, the mixer will have to ask the prop person to attend to the fittings so they don't make any noise.

But suppose that in editing, the director decides that the sign made such a distinctive sound that he would like to use a tight shot of it to open the scene. If the mixer didn't have the foresight to make some wild tracks of the sound before it was altered, it will be the job of the sound effects editor to come up with other squeaking sounds that may or may not match the original sound. Of course, once the director has a particular sound in mind, it is unlikely he will settle for a substitute. Thus the beauty of wild tracking.

Room tone. Whether the term used is *room tone, ambience,* or *extraneous noises,* the resultant sound is vital if a film is to be edited successfully. Although the term *room tone* normally is associated with interiors, many directors will use it to describe general atmospheric sounds on exterior locations. Room tone is the overall sound that makes a particular location unique. Unlike wild tracks, it is not a separate sound but a combination of many subtle sounds. In the case of our diner scene, it might be the sounds created by a slight breeze blowing or faint traffic sounds heard in the distance. In short, room tone is affected by many things—atmospheric conditions, a slight change in location, even the position of the microphone. Consequently, a sampling of room tone must be made for editing purposes *at that point in time the scene is shot.*

Suppose scene no. 1 was shot at 8:30 P.M., but it wasn't until 1:30 A.M., with the conclusion of take no. 24, that the director finally wrapped (ended) the scene.

As it turns out, the director likes take no. 1 for the beginning, take no. 12 for the middle, and take no. 24 for the ending. From the time the shooting began at 8.30 P.M. until it ended at 1.30 A.M., the ambient room tone changed. However, because the mixer sampled the room tone sound after each take, the director has a choice of which one room tone will be played through the entire scene. Had the mixer not done his job, each of the three portions would have a slightly different ambience, making each edit noticeable.

To cover the edit changes in sound, background sound effects of wind, crickets, and distant traffic can be added. If an edit point still has too much dropout (dramatic decrease in sound level), the discreet use of a dog barking or some other appropriate effect will mask (cover) the edit point.

Be it in the studio or on a remote location, although it is never stated in so many words, everyone knows that pictures take precedence over sound. This means the production mixer's main concern is dialogue and ambient room tone. The best the sound editor can hope for is that the tape of ambient room tone will give a clue for the background sounds, and just hopefully, the mixer will wild track such unusual sounds as that huge swarm of killer bees attacking a ferocious leopard. But then again . . . would you?

The Boom Operator

The boom operator and cable people, who work with the mixer, operate the dialogue boom mikes and fishpole mikes, and attend to the tape machines. Compounding their job, they must keep out of the camera operator's shot, avoid dolly tracks, stay out of the gaffer's (electrician's) light, not hinder an actor's movements, and remain inconspicuous. Furthermore, they must see to it that the mike is in the best possible position for good dialogue sound.

Cable People

Cable people lay cables for sound equipment, move the equipment for shot changes, reroute cables, and generally make themselves available to assist the mixer and boom operator.

Sound crew members are also responsible for the placement of radio frequency microphones in the clothing of the actors, in effect, "wiring the actors for sound." These tiny microphones contain no wires and free the actors from the limited movement imposed by the boom microphone.

THE SOUND EDITORS

The sound editor's job is to add effects to a film. They rarely call them *sound effects* because the word *sound* in film is such an all-encompassing one. Therefore, sound in film is referred to as dialogue, music, and effects.

Suppose in our biker film, the local police arrest what they believe to be all of the gang members; however, one escapes and telephones for help from other gang members.

To intensify the excitement and tension, the director decides to show the gang racing to the rescue using tight, medium, and long shots in sync with music. Although this is a rather complicated mixing job, it can be simplified considerably.

Because the sound effects tracks illustrated in Figure 9–1 have been edited precisely to correspond with the picture, all that remains to be done by the mixer in post production is to set satisfactory loudness levels on the three tracks with assistance from the director. Now when the transfer of sound effects is done to match the film, the mixer opens the proper fader according to the footage numbers on the cue sheet and the loudness levels are in the correct perspective.

However, notice in Figure 9–1 that when the fader is open on sound effects track 3, the effects are heard through the entire scene. This is permissible. Because this is the track containing the motorcycle sounds at a great distance, their loudness level will be masked by the loudness of the motorcycle sounds from the other two tracks. Therefore, this track needs no blank tape leader and may be left playing throughout the entire scene.

This technique has many applications, but whenever you are playing background sounds and you want to avoid a "choppiness" of abrupt level changes, layer the sounds in such a manner that they all do not go out at once. For instance, if a scene calls for a character to start in loud traffic and enter a quiet apartment, layer one traffic track at the perspective you want when the apartment door is closed and have that play under your other

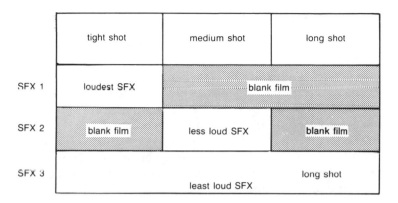

FIGURE 9–1 The top portion of this drawing represents the picture portion of the motorcycle gang shot from three different perspectives. The lower portion represents the sound effects tracks with three different loudness levels. On sound effects (SFX) track 1, the sounds are the loudest because we show the bikers close up in a tight shot. On SFX track 2, the sounds are less loud; and on SFX track 3, they are least loud because the camera is taking a long shot.

traffic tracks. When the door closes, you fade out the louder tracks that have been masking your low-level track and keep the off-perspective track in until such time as you want to gradually fade it out.

SUCCESSFUL SOUND EFFECTS EDITING

The most important point to remember, regardless of whether you are doing films or television, is that all successful editing is a matter of control, control, control. Although it often seems like an awful waste of time and energy on the part of the editors, effects for a scene must be as isolated from each other as possible so that the producer and director have the freedom to say what effects they want to hear in a scene. This method, especially in films, is time-consuming and therefore expensive, but examine the alternative. The effects artist assembles all the sounds on a composite track according to how she interprets the scene. The director hears the track and is in total agreement except for "a few" changes. "Take out the dog bark, lower the level of the crickets, reduce the number of cars driving past, and raise the level of the motorcycle screeching to a halt." Out of a scene containing perhaps fifty different sounds, with fifty different levels, those changes are indeed only "a few." But whether they are just a few or only one, when dealing with a composite track where all sounds are mixed together, no individual sounds can be changed or removed.

Syncing Sound Tracks

The picture and sound on all major films are shot separately. Prior to each take, the assistant director holds up a slate/clapstick and reads the information written on the slate pertaining to the scene and take number. Then she strikes the clapstick, and the scene begins. Each time the scene is repeated, the slate is revised to reflect that particular take.

While the camera shoots the scene, a separate $\frac{1}{4}$-inch tape recorder runs in sync with the picture. After the scene has been shot successfully, the director decides what scenes are to be printed. This is noted on a camera and sound report so that later the editor will have the correct information for editing purposes.

Cleaning Tracks

Post production sound is divided into three categories: dialogue, sound effects, and music. As mentioned previously, the most important consideration given to sound is the dialogue. Therefore, when the dialogue editor receives the production sound tracks, the most important job is to "clean the tracks" of all sounds other than dialogue. This entails editing out all spurious sounds and replacing them with room tone. When an unwanted sound such as a car horn is edited out, it must be replaced with an equal amount of room tone. If this isn't done and the edit point is simply joined by a piece of blank film, there will be a noticeable dropout of background ambient sound.

Incidentally, the information gleaned from these unwanted production tracks, referred to as "scratch tracks," is often extremely valuable to the sound effects editor. Although they may be unusable because of various noises such as a director shouting instructions or their assistants screaming for quiet, the background sounds will give the editor an idea of what sounds are needed for that particular area. If the action takes place aboard a cruise ship anchored in a harbor, the editor will be able to tell how busy the port is and what types of boat whistles can be heard.

Bouncing Tracks

Very often a scene will require as many as fifty different reels of sound effects tracks. Playing all of these reels at the same time would be extremely difficult, even for the most sophisticated post production houses. As an alternative, there is a technique referred to as "bouncing tracks," which involves reducing the number of sound effects tracks that are similar in nature by dubbing (copying) two or more tracks onto a single track.

For instance, in our biker film, there may be five sound effects tracks involving background cricket sounds. Although each of the tracks is slightly different from the other cricket tracks, it is fairly safe to say that these five tracks could be mixed together to make one cricket track. The same would be true of three slightly different gentle wind tracks.

Very often the decision of whether several tracks can be bounced together is a matter of how similar the sounds are and what level of importance the effects have in a scene. In the cases of the crickets and the wind, there should be no problem. But suppose the editor, in an effort to reduce the number of sound effects tracks, adds a dog barking to the cricket tracks. Although the editor was careful to keep the level of the bark in proper relationship with the loudness of the crickets, the director likes the dog bark and wants to hear it louder. This is one of the dangers of bouncing tracks. Because the dog bark is on the same track as the crickets, making it louder also increases the sound of the crickets. Unfortunately, this renders the track unacceptable, and it must be made over. To avoid this problem, dub together only those tracks that are so similar in nature that if the loudness level must be raised or lowered, it won't make any part of the overall sound stand out.

The other consideration in bouncing tracks is the importance of that sound to the scene. Gunshots are a good example, because everyone has their own idea about how gunshots should sound, especially if the gunshots command a high degree of focus in a scene.

To attain just the right gunshot sound very often requires as many as twelve or more tracks. On one reel there might be three frames from a clap of thunder; on another reel, four frames of a lion's snarl. And on the twelfth reel there might be the sound of an old shotgun fired in a Foleying room and recorded by a 416 Sennheiser mike, wrapped in a rubber glove and submerged in a pail of water to give the shotgun a dull, nonreverberant sound. But there will be other recordings too: the sound of a gun firing a live round into a bag of sand, that of a gun firing a blank at the ceiling, into a barrel, in the lavatory, and any number of other combinations.

When they are all mixed together (but still on separate reels), the director can go to a dubbing stage and listen to the mix of the tracks to determine how it matches the gun in the scene. One of the most difficult tasks is to listen to a sound and imagine how it will match the picture. Just the opposite of radio effects, where the sound suggests the picture, in film and television, the sound must support or enhance what is seen.

When the director is happy with the various combinations of sounds, all the tracks can be bounced to one composite track. Be forewarned: to bounce these effects down to one track or even to several tracks without the director first approving the sound itself is asking for trouble.

Bouncing tracks should be done only to reduce the tracks with similar sounds down to a more manageable number. It should never be done at the risk of losing control over individual sounds.

LOCATION SOUNDS

Excluding the elements of time and budget, films and television shows shot on location have many similarities. The primary concern is always the pictures, followed by dialogue; if there is time left over, it is used to get more pictures. Sound effects are given little, if any, consideration. The reason for this is quite simple. Producers have found out that by excluding as many natural sounds as possible, they are better able to control the mood and tempo of their film or tape by adding only the sounds they want during post production.

What Sounds To Record?

Although modern sound effect libraries have thousands of unusual sounds, it is impossible to have just the right sound for every situation. What determines whether or not a sound you want is in a library? Sounds are recorded and kept if there is some likelihood they might be used again or if the sounds are so unusual that the chances of ever having the opportunity to record them again live are extremely remote. The *Ben Hur* chariot race is an excellent example. No television or film company is ever going to stage a chariot race for the sole purpose of recording its sound. Therefore, the opportunity to record a sound so extraordinary as a chariot race is almost never refused. I say *almost* because the sound effects artist is rarely present at a remote location. What sounds are recorded is usually left up to the production sound mixer and the director. But again, such a scene, which costs millions of dollars, is not going to be staged simply for the benefit of its sound. In fact, it is probably safe to say that the sound of the race is the very last thing the director is thinking about. Therefore, even if it is being recorded, there will also be the sound of helicopters overhead taking aerial shots, the stunt coordinator shouting instructions, and the director screaming obscenities. Through this din we might just hear bits and pieces of a chariot race.

Obviously, these sounds do not go into the library file. What is put in the library are those bits and pieces edited out of the overall production sound track and expanded upon to match the finished, edited film—complete with thundering hoofbeats, whip cracks, chariot sounds, and appropriate crowd reactions.

If your particular *Ben Hur* involves seven semitrailer trucks racing across a desert, the same considerations hold true. There will be the same helicopter sounds, shouting, and cursing, making it impossible to hear the truck sounds in the clear (without other sounds). Similarly, bits and pieces of truck sounds will have to be expanded upon to create the sound of trucks racing. But suppose there are two people in each truck and there is dialogue between them. Unless this dialogue is going to be "looped" (re-done in the studio), the sounds of the various trucks heard on the original dialogue track are going to have to be matched.

This does not mean the sound of the trucks racing has to be recorded in its entirety, but it does necessitate wild tracking (nonsynchronous recording) sounds that most closely match those made when the dialogue was being spoken. The reason for this is that two or more sounds, once they are recorded together, can never be separated. Never ever! Therefore, when dialogue is spoken at the same time that a sound is occurring, the two are "married" for life. The options are looping or wild tracking that particular sound so the artist can match it in post production. If the sound in the background during dialogue is loud enough to be heard and yet has nothing to do with the scene, one other option exists.

Suppose a highly emotional scene is shot (which everyone loves) and it is discovered there is the unmistakable sound of the camera's generator humming in the background. Rather than reshoot the scene (which the actors refuse to do), the noise can be *masked*, or covered with another (louder) sound. Of course, the sound we are masking with, no matter what it is, must have a reason for being there. And that is where the director must justify the sound with an added shot.

For instance, if the scene takes place in a seedy motel, we can establish the sound of nearby traffic; or if it's hot outside, one of the characters can turn on an air conditioner. Anything can be used as long as the sound source is established. As soon as we're into the scene, the effect that we're using for masking can be slowly faded until we're back to the original generator hum, only now no one in the audience will be aware of it.

MANIPULATING REALITY

When Orson Welles manipulated the reality of his radio show *The War of the Worlds* by presenting it as an actual news program, he caused a public furor, the ramifications of which can be felt today. It is now against Federal Communication Commission rules to present a program as an actual newscast simply for the dramatic impact it creates. Yet, in a much less serious way, films and television are constantly manipulating our perceptions of reality.

When a film opens on a scene depicting waving green pastures, grazing cattle, and distant mountains, we become absorbed with the pastoral beauty. And when there are accompanying sounds of lowing cattle, the scene's tranquility is complete. However, as the camera begins pulling back, we notice the bucolic sounds are beginning to slowly fade. Suddenly, our scene is framed by an octagon ring. As the camera continues to pull back, we find our scene enmeshed in a wire fence; what was perceived as spatial expensiveness was an illusion created by the camera shooting through the fence at close range. As we hear the sound of an off-camera siren, we are further confused. Then the camera once again widens its angle; we realize the scene we have been viewing is shot from the perspective of a prisoner behind a high barbed-wire fence. Our perception of reality has been manipulated by the camera's angle.

SPATIAL CONFINEMENT

Just as our perception of spatial reality was colored by the camera's angle, sound effects can enhance the reality of a scene.

Filmmakers have learned that an ordinary story suddenly becomes believable and interesting when shot on location. This also gives the director a great deal of spatial freedom. However, cameras can only be focused on one object at a time. This is where sound effects come into play; sounds can free the audience of the screen's spatial confinements. Furthermore, they can confuse the viewer's sense of reality by promising sound sources that do not exist. If the camera only shows five cattle grazing in the opening of the prison scene, the added sound of cattle lowing from a greater distance suggests a larger herd. When the siren interrupts the scene and the viewer is shown a guard tower, the natural assumption is that the sound is coming from that location, when in reality, the siren might be located in the laundry building on the other side of the yard.

Utilizing off-camera sounds for spatial purposes is not unlike the function that sound effects served in radio. Although an audience is most influenced by what it sees on the screen, the use of off-camera sounds helps to overcome a scene's spatial confinement.

Obviously, sound effects should not be compartmentalized. If you understand the proper use of sounds in one medium, you can apply this knowledge to the other media. The only differences are the techniques, and the equipment employed.

SUMMARY

1. To attract audiences away from television, companies spend a great deal of time and money producing films. Consequently, sound effects receive more attention in film than in television.

2. Although the number of people directly connected with the picture part of the film may run into the hundreds, very often the crew responsible for the sound numbers as few as three.

3. The picture and sound on a feature film are separate.

4. Very often the film that the dialogue editor discards when cleaning the production tracks is extremely useful to the effects editor in deciding what ambient sounds to use.

5. Normally, the effects editor and the composer will work together closely to determine whether music or effects will be featured.

6. To give the sound mixer good control over various effects, the sound editor will allow a distance of five feet between effects. This is the amount of time the mixer needs to open and close a pot.

7. The picture and sound portions of a film are synced up with the slating clapboard.

8. Sound cue sheets are a log describing where sounds can be found on a track.

9. A reel of film is 1,000 feet in length. The first twelve feet are used to bring the film up to proper speed in the projector. This portion is referred to as the "academy leader."

10. In addition to film footages, film is broken down further into frames. There are sixteen frames in each foot.

11. During a spotting session the director and the effects editor view the film and decide what effects are needed.

12. A streamer is a two-foot-long mark drawn on the film with a grease pencil to alert the editor of a looped effect.

13. All editing, whether it is for television or films, is based on giving the editor control over the effects.

14. Reducing the number of effects tracks can be accomplished by bouncing, or moving effects to a composite track of similar effects.

15. The dubbing stage reduces the number of effects reels.

16. One of the major differences between television and films is the attention that films pay to the effects that television either ignores or covers with music.

CHAPTER 10

FOLEYING

Despite the sophistication of modern sound effects equipment, much still depends on the artist. He must coax the desired sounds from inanimate objects using his hands and sometime even his feet. In other words, it's just like the old days—with one exception. In the beginning, the artists used manual techniques out of desperation, whereas today, they do so by choice.

At this point, we should clear up a misconception about the term *Foleying*, which is derived from Jack Foley, a sound editor for many years with Universal Studios. In 1950, while working on a film titled *Smuggler's Island*, starring Jeff Chandler, Jack Foley had to lay in the sounds of Chandler paddling a small survival raft in the Pacific Ocean. Foley decided to do the ocean and paddling sounds live in the sound studio rather than use the more conventional method of cutting the effects into the film itself. The efficiency of this technique impressed many people, and soon other directors wanted to implement sound effects the way "Jack Foley did it." Incidentally, if you are ever doing a film in Mexico, they don't refer to syncing effects as *Foleying*; in Mexico the term is *Gavrila*. When they want an effect synced live in Mexico, they want it done like "Gavrila does it."

In fairness to all the other effects artists in film, Jack Foley did not invent this technique. His name simply became identified with it. Prior to this, sounds that needed to be laid down on film were referred to as *synchronized effects*. At Paramount Pictures, for instance, the term was *make and sync*. This meant doing the effects on a sound stage into an open mike in sync with the action of the picture. The techniques, and very often the effects themselves, are exactly the same whether performed manually by artists for radio and television or by Foley artists for films.

To give you some idea of the extent to which some Foley artists will go to achieve a desired effect, Howard Beals, a sound editor at Paramount for many years, worked on all of the films done by Cecile B. DeMille at Paramount. For the effect of the Israelites making bricks in the film *The Ten Commandments*, two Foley artists worked in a large tank of mud and water. But because their clothing made a rustling sound, Beals, at DeMille's insistence, had the artists simulate the brickmaking in the nude! And how did your day go at the office, dear? (See Figure 10–1.)

SOME FUNDAMENTALS

Because a Foleying stage in a post production house will not generate the profits the other equipment will, it is unrealistic to expect an elaborately designed room just for Foleying. Even at the large film studios, Foley stages often appear to be storage areas for the studio's unwanted junk. Metal laundry tubs are filled to the brim with metal trays, tin pie plates, empty soda cans, hubcaps, bedpans, knives, forks, spoons, metal bookends, and broken staple guns. These *crash tubs* are used for anything from comedy crashes to adding presence (brightness and naturalness) to something as serious as a car crash. In one old battered roll-top desk, drawers are filled with old shoes: sandals, wooden clods, bedroom slippers, tennis shoes, high-heeled shoes, flats with metal taps, moccasins, galoshes, and cowboy boots—if people walk in them, you'll find them in one of the drawers. The desk, in addition to being a functional storage area for shoes and smaller props, is useful for its own sound of a roll-top desk. Embedded in the floor is the heart of any Foley stage—the walking surfaces. No matter how small your space or how modest your Foleying needs, provisions must be made for the production of all types of footsteps.

Figure 10–2 illustrates some of the surfaces needed to produce the sounds of footsteps. Additionally, you will need a small marble slab and linoleum tiles glued to a thick plywood board. For scripts that call for steps in New England's woods or an African jungle, dried leaves and small bunches of hay worked with the hands will give a very realistic effect. (See Figure 10–2.)

Doing Footsteps

For a woman to produce the sounds of a man's footsteps, she needs a pair of sturdy low-heeled leather shoes. For a man to do a woman walking, he need not purchase an oversized pair of high-heel shoes. Rather, he

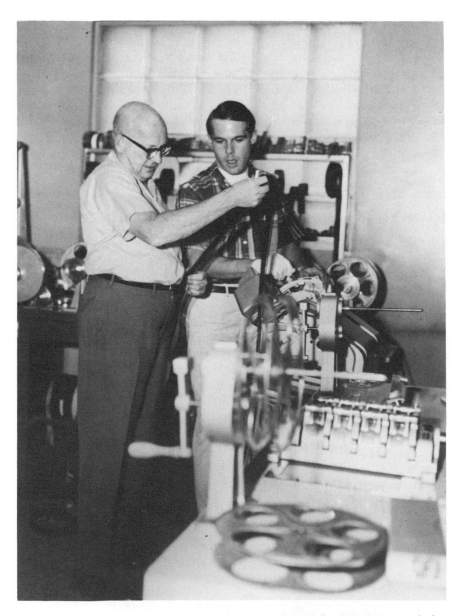

FIGURE 10–1 Howard Beals, a sound editor at Paramount for many years, worked on most of the Cecil B. DeMille films. Although here he is inspecting a piece of film that needs sounds from Paramount's huge library of effects, very often, to satisfy the exacting demands of DeMille, Howard found himself out on the back lot creating the sounds from scratch. (Photograph courtesy of Donna Beals.)

FIGURE 10–2 Dianne Marshall, a free-lance Foley artist, syncs footsteps to a scene being projected on a screen. Notice the various surfaces designed to accommodate the artist during a busy scene. While the steps may begin on a concrete sidewalk, during the course of a scene they may move to grass, gravel, and wooden stairs (with or without squeaks).

can simply walk on the back of his heels. By using just a small area of the heel (and not his full weight), this sound is quite convincing. More important than the sound itself, to be truly effective, the man must capture the rhythm of a woman's walk. Without this vital sound component, his efforts will be less than satisfactory.

Walking In Place

Another reason why the area allocated for Foleying doesn't have to be large is that all the effects are normally done around one microphone. In order for the footsteps to be clean and crisp, they are done in place. This is what makes doing steps so difficult. It is one thing to "march in place," so to speak, but quite another to do so without it sounding exactly like you are doing just that. Although you don't have to be a dancer to do steps, you must be enough of an actor to play the part of the person you are walking for—regardless of their sex. This was true of radio and it is true of films and video. In order to be a good Foley artist, you must become the character on the screen with whom you are syncing your effects. If you just lift one foot up and put it down in sync with the picture, your steps will sound mechanical and robotlike. However, once you become the character you are Foleying for, you'll be amazed at how natural the sounds become.

MIKING THE SOUNDS

One of the best pieces of advice that can be given regarding microphones is: Don't use more than you have to. For Foleying purposes, this means one mike. Obviously, that one mike must be conveniently mounted so that it will be readily available to a number of different positions. The most common method of hanging a mike is either with a baby boom or a goose neck attachment. In this way, the mike can be lowered, raised, extended, or retracted with relative ease.

If you are working in a comparatively small room with a lot of other equipment, it would be a good idea to get a baffle (a portable acoustical screen) to isolate the area you are working in from the rest of the room. These baffles come in various models. Some have one side that is reflective to sound and one that is absorbent; while others have both sides either reflective or absorbent.

Mike Techniques

The microphone you use depends on the bulk of your work. Most high-quality mikes will do the job satisfactorily if the room is quiet. If you have a lot of equipment noises, you will probably want to consider a more directional mike such as a hypercardioid microphone.

How close or far away you place the mike from the effects is too subjective a matter to advise. Simply do what all professionals do: Start off with the mike one foot away and move it closer or farther away until you get the desired sound, and do enough takes to ensure a wide selection. This is especially important when you are at a remote location. Although it is extremely important that you have the proper microphone, it is even more important that you use it creatively.

Contact Microphones

Unlike standard microphones, a contact mike is attached to an object and depends on the vibrations of the object for its electrical energy. This microphone was used for recording sound effects in the earliest days of radio (see Figure 1–3). Like all microphones used in sound effects, contact mikes are selected for their excellence over a wide range of demands. Many conditions will only be encountered once; consequently, you must have equipment that will most safeguard success. Because contact mikes are attached directly to an object, interference from other sources is eliminated.

Shotgun Mikes

So called because of its elongated interference tube that blocks out extraneous noises, a shotgun mike is highly directional and extremely useful when trying to focus in on one sound in an otherwise noisy environment.

Pressure Zone Microphones

The pressure zone microphone (PZM; trademark of Crown International Inc.) is unconventional in appearance and has a somewhat limited application. The transducer is mounted on a flat surface with a $\frac{1}{32}$-inch separation between the transducer cover and the plate. The main advantage of this type of microphone is that the direct and reflected sound normally associated with a conventional mike is practically eliminated with a PZM mike.

WHY FOLEYING?

Although Foleying is far from a precise science, it can impart touches to a scene that either are overlooked during the production shoot or would be extremely time-consuming and expensive with more conventional methods. Let's look at an example of a typical scene being shot on location.

The scene is a small sheriff's office. Outside the window we can see the wind blowing clouds of dust down the street. Through a doorway we can see a jail cell. The sheriff, with his back to us, has just led the prisoner into the cell and slammed the iron-barred door and is in the process of locking it.

> PRISONER:
> (TAUNTING) What do you figure, Sheriff ... an hour ... maybe a little more with this dust kicking up like it is ... but for the time it takes my brothers to get here is just about all the time you have to live.

The sheriff finishes locking the jail door, walks into the office, and closes the door leading to the jail. He hangs up his keys, looks up at the wall clock, goes to the desk, unlocks the drawer, takes out another pistol, opens it, and begins loading it with bullets from his gun belt.

Assuming there will be no music added to this scene, you are dealing with three types of sound effects: those that are on the production track, those that will be cut in, and those that can be done on the Foleying stage.

Let us review our little scene and try to decide what effects are involved and how they should be created.

1. Dust storm

2. Jail gate

3. Jailer's keys

4. Boot steps (spurs?), wooden floor

5. Keys hanging up on nail

6. Door closing

7. Wall clock ticking

8. Unlocking desk drawer

9. Opening desk drawer

10. Pistol out of drawer

11. Opening pistol's cylinder

12. Removing bullets from gun belt

13. Rustling clothes

14. Inserting bullets into gun

This list illustrates the versatility and convenience of Foleying. With the exception of the dust storm, the clock ticking, and possibly the doors, these effects can be done on the Foleying stage. Translate this into the time saved in editing, multiply it by the many scenes in a feature film that require effects, and the advantages of Foleying becomes obvious.

THE FOLEY STAGE

Although the size of the Foleying room may vary from television to films, they perform exactly the same function—to help the artist synchronize manual effects with the actions being done on tape or film. Interestingly enough, the Foleying stage for television is considerably smaller than that for film. This is also true of the monitoring screens the two artists watch. In television, it is the size of a 21-inch television set; on the film stage, it is just slightly smaller than the size of the screen at your local theater. Although the television artist would benefit equally from the larger screen, the producers in each of the two fields prefer to listen to the sounds in the context in which they will ultimately be heard. This feeling of "seeing what I'm going to hear" is more than just a psychological quirk on the part of the producers. In films, because they are projected onto such a large screen, things that are unimportant or go unnoticed in television are very apparent in films.

If, for instance, you are doing the identical gunfight scene for a feature film and a television program, the number of sounds used and the emphasis placed on these sounds will vary greatly. Although this will also vary from film to film according to sound effects budgets, even films with relatively low budgets will give certain scenes, certainly the close-ups, particular attention to sound.

Going back to our western scene, to be used in a film, if the camera were tight on the sheriff's holster as he strapped it to his leg, the action would demand the rustle of clothing and small squeaks of leather. And when he practiced drawing his gun from its holster, this sound, plus the clothing sounds, the scrape of his boot on the wooden floor, the wall clock ticking, and perhaps a pot of coffee perking on the potbelly stove, would fill the room and point up the sheriff's isolation at a time of danger. This would all work because of the immediacy an audience feels watching these huge figures projected on the screen.

Perhaps the most advantageous reason for doing effects live into a microphone is the clarity of the resultant sound. When sound effects bypass all audio equipment other than the mike, they are as close to the original sound as possible. These sounds are first-generation quality. If taped, they would become second-generation sounds. Each step down in generation results in a loss in fidelity, in much the same manner that a picture can be repeatedly enlarged to a point of indistinction.

As digital recorders become more accessible, the generation loss problems inherent with analog recorders will no longer be a consideration. This, however, will not make the Foleying technique any less attractive. The controllable speed and efficiency of syncing sounds live will always have a place in films and television.

SOME PRECAUTIONS

Although Foleying is a tremendous aid, it should not be looked upon as a panacea for sloppy production values. One of the biggest drawbacks to Foleying is that it is done in a soundproof studio. As a result, unless a great deal of attention is given to the sound ambience, the Foley effects will not match the sound quality of the film.

This is especially true of footsteps. Very often, after the director has shot a scene of a man hurriedly walking across a lonely street at night, she feels her job is done. She doesn't have to give any attention to production sound because she knows a Foley artist will be able to "lay in" the steps in perfect sync. Thus the matter is dismissed until post production.

When the sweetened scene is played back, we have a man hurrying across the lonely street, footsteps clicking along in perfect sync, with nothing to break up their sound but (hopefully) the street's ambience. The problem is, unless this ambience is extremely interesting and the scene is not too long, the clean-sounding sound-stage footsteps will begin to dominate; and unfortunately, the scene has been shot and there is nothing that justifies sounds. This is when the director begins pulling in sounds from left field. (As witness, how often you have heard the ubiquitous television and film off camera dog barks?)

To avoid this total reliance on Foleying to carry a scene, the director should select a location not just for its eye appeal, but with a thought toward how the scene will sound.

IN CONCLUSION

In all our discussions, there is one sound we have totally ignored—silence. The absence of sound is an alternative to sound. To the sound effects artist constantly searching for new methods to present sound, it is another option.

In early radio, sponsors dreaded it. So too do young couples on their first date. To a soldier in battle, "things can be even too quiet." City dwellers often prefer the traffic screeches and sirens to the "nothingness" of the country.

In each of these cases, it isn't the silence that is so objectionable, but rather what it represents. In early radio, the absence of talking was referred to as *dead air*. Likewise when an actor is late with his cue or takes too long to deliver a line. It is this very preoccupation that everyone—not just the people who make films or television shows—must have some kind of voice going all the time that gives silence its tremendous impact.

If, for instance you were the sound designer for this scene, how would you handle the sounds to make them unusual? The scene takes place in a busy, noisy restaurant where the family patriarch is celebrating his 75th birthday. There is music, dancing, food, drink, and laughter when suddenly the grand old man suffers a heart attack. As he falls against the table, dishes and bottles crash to the floor, a woman screams, and the gasping man falls heavily to the floor dying.

Aside from the commotion from the restaurant guests and family members, the woman's scream, dishes and bottles breaking, and the man's body fall, what one thing would you do to make the scene more compelling and terrifying?

One possibility would be to shoot the scene from two points of view—the family members' and the victim's. Prior to the heart attack, there are sounds of happiness—music, talking, dancing, and laughter. When the attack occurs, we cut to the victim's point of view and all restaurant sounds are suddenly cut. Not even the dishes breaking on the floor make a sound. The only sound we hear on the victims side is silence punctuated by hoarse wheezing and gagging. As we cut back and forth, the two sounds are counterpointed. Therefore, when death comes to the old patriarch, it isn't accompanied by darkness, but the one sound most of us dread—utter and complete silence—the absence of what constitutes life . . . sound.

SUMMARY

1. *Foleying* is a film term used to describe synchronous effects or live effects. It literally means doing the effects manually in synchronization with a picture.

2. Foleying is an excellent means of supplying the subtle sounds that production mikes often miss. The rustling of clothing and squeak of a saddle when a rider mounts his horse give a scene a touch of realism that is difficult to provide using other effects methods.

3. The good Foley artist must "become" the actor with whom they are syncing effects or the sounds will lack the necessary realism to be convincing.

4. Most Foley artists use a hypercardioid condenser microphone for their effects.

5. Most successful Foley artists are audiles; they can look at an object and imagine what type of sound it can be made to produce.

6. Because Foleying is done live into an open mike, resultant sounds are first-generation.

7. Because of the clarity of first-generation sound, sometimes matching the Foley sound with the production sound can be troublesome.

8. Despite the versatility of the various methods of laying effects down on film or tape, nothing can compare with the speed, convenience, and versatility of Foleying.

GLOSSARY

A-B'ing   The term used to signify that two separate sources are involved. Comparing two sources by switching from one to the other.

Acoustics   The science of the behavior and control of sound. Although this is a highly technical field, the acoustics of any enclosure is that particular quality that allows a good "sound" for your particular needs.

Ad libbing   Anything that is spontaneous and unrehearsed.

Air   The word used to designate the actual show that will be broadcast.

Ambience   This pertains to the pervading atmosphere of a place. More of a psychological, rather than technical description. A restaurant may have a warm, friendly "ambience" for its patrons, whereas, the pianist may very well be critical of its "acoustics."

Amplifier   A device used to increase the level of a sound signal.

Amplitude   The strength of a sound signal.

Atmosphere people   Performers without speaking lines. This term has replaced "extras." *See* U5's.

Attack   The manner in which any sound begins. A clap of thunder has a strong "attack" . . . the soft rustling of leaves has very little.

Attenuator   A device used to control sound levels. *See* Potentiometer; faders.

Audition   Performing a task for the explicit purpose of seeking approval; *also,* a system of listening without feeding the line

Backtiming   Normal measurements of time begin with zero elapsed time; backtiming begins at a certain point of elapsed time and measures back to zero.

Baffles   Objects composed of sound-deadening materials and used to isolate voices, musical instruments, or sound effects in a studio.

Balance   Relative levels between different sources.

Bandwidth   The range of lower and upper frequencies.

Beat   Duration of approximately 1 second.

BG   Background.

Bidirectional microphone   A mike that picks up sounds from two sides, the front and back, but is dead on the sides. Some of the older mikes, such as the RCA 77, have an adjustment screw that allows you to change their pattern.

Biz   *See* Business.

Black   "Absence of a video signal." Television writers use the absence of a video signal to indicate the beginning or end of a scene. At the beginning of a scene, it is normally written simply, "FADE IN.", whereas to indicate the end, it is, "FADE TO:BLACK."

Bleeding   A serious video problem involving color signals crossing into adjacent images.

Blocking   Planning the moves of actors and technical equipment.

Blooping   A method of erasing short portions of tape.

Board   Audio mixing console.

BOC   Broadcast Operation Center.

Boom   Most often refers to a piece of audio equipment that contains a platform for the operator to stand on and a retractable steel arm mounted on three wheels. The microphone is attached to the end of the arm and may be raised, lowered, or swiveled in a 360-degree arc by the operator.

Boost   Raise the sound level.

Bounce   The amount of sound wave reflection off a surface.

Bouncing tracks   When using a multitrack recorder, the need for additional channels can be solved by transferring, or "bouncing," information from other channels and mixing them down to one channel.

Brilliance   The treble quality of an audio signal.

Bump   Raise the level.

Burning tape   Unnecessary waste of tape or erasing tape.

Burst   Briefly feeding an audio signal for a test of continuity.

Bus   The output line of a mixing network.

Business   Working with props. It can be as simple as putting cream in coffee or as elaborate as setting a dining room table for a Thanksgiving dinner.

Busy   Too many things happening at once.

Butting   Getting as close together as possible.

Button   Bring a musical number to an end.

Buy The word used for acceptance.

Calibrate Balancing various pieces of equipment so they all have similar standards.

Camera blocking The first rehearsal utilizing technical equipment.

Camera left To the left of the camera from the viewpoint of the camera person.

Camera right To the right of the camera person.

Canned Prerecorded music or audience reaction.

Cans *See* Headset.

Capstan The shaft that rotates and feeds the tape prior to the head. When you have an extremely tight cue, it helps to rapidly rotate the capstan just before the cue.

Cartoon effect An exaggerated sound effect in the fashion of a cartoon, for example, boings, slide whistles, whizz bangs, and temple blocks. All are totally inappropriate on a dramatic show.

Cartridge The housing for a continuous loop of lubricated tape; *also*, the transducing element in a record player's stylus.

Channel A single sound path.

Control room The brain center where the audio and video levels are controlled.

Control tracks These contain the pulses that synchronize video playback and the beginning and end to each frame.

Cough switch Same as mute switch.

Countdown A 5-second ready cue.

Crabbing Moving the boom mike in a left-to-right or right-to-left fashion.

Cropping Cutting off part of the video image.

Cross-fade The technique of fading in a new sound, as you fade out an old sound, without any noticeable difference.

Crosstalk In multichannel audio systems, a signal leaking to another channel.

Cue A signal to start. There are three different types of cues that relate to sound effects. A verbal cue comes from the control room; a word cue comes from the performers; and a sight, or business, cue comes from a visible action.

Cue track Portion of a video tape track that records information vital to editing.

Cut (1) In video or audio, going from one picture or sound to another, abruptly and without loss of level. (2) On records, the separate sections of music or sound effects, also called *bands*. (3) *Cut* The command given to stop all action in the studio.

Cyclorama Cloth or canvas material the height of a studio used for background.

Dead pot   Rolling a tape or record at a specific time without opening the fader.

Decay   The end portion of a sound.

Decay time   The measurable time it takes a sound to fade.

Decibels (dB)   One decibel equals .775 volts root-mean-square.

Degauzer   An instrument for erasing audio and video tapes. (And ruining your wristwatch if you forget to take it off!)

Digging   Raising the gain of a microphone in an effort to pick up low-level sounds.

Distortion   A signal too loud for the playback system.

Dolly   (1) Physically moving the camera closer or pulling away from a subject. (2) A low four-wheel cart.

Doppler   The perceived increase and decrease of a frequency source as it passes one point.

Dramatic license   Sacrificing the reality of a situation for a desired dramatic effect.

Drop out   A drastic loss or drop in level due to an unevenness in the tape coating.

Dry block   A rehearsal in television without the cameras or microphones being operational. In radio this is referred to as a "reading," or "read-through," while in films there is no particular term other than simply "rehearsal." *See* Dry rehearsal.

Dry rehearsal   Running scenes without technical facilities.

Drying up   Unable to speak; "mike fright."

Dubbing   Transferring audio or video from one playback source to another.

Dynamic range   A span between the loudest and softest sound a source can produce before distortion occurs.

Echo   The repetition of sound caused by early reflections. *See* Reverberation.

Echo chamber   Created by placing a loudspeaker and a microphone in a room with highly reflective walls and floor. As the signal from the studio comes out of the loudspeaker and bounces off the different surfaces, the microphone picks up these reflections and feeds them back to the studio in the form of *echo.*

Energy   Pertains to the enthusiasm that actors put into their performances.

Equalizing   Adjusting the frequency responses for a desired sound quality.

Erase head   Electromagnetic transducer that demagnetizes the tape. *See* Degauzer.

Eyeballing   Applies primarily to sweetenings and post production work; it means taking your cues from the monitor or screen as opposed to the time code numbers.

Fade in   A smooth graduation of signal from *black*.

Fade out   Smoothly decreasing the signal to *black*.

Fader   *See* Potentiometer.

FAX   A television term meaning "facilities." When a rehearsal is scheduled with FAX it means a rehearsal with all facilities, including lighting, scenery, props, cameras, sound, sound effects, and special effects (other than elaborate and costly effects such as explosion or fires that may be saved just for air). Makeup, hair styling, and costumes are not normally seen until dress rehearsal and air.

Feed   *See* Signal.

Feedback   A high-pitched squeal caused by information from a speaker feeding back into an open mike.

Fibbers' crash   A long crash produced by pots and pans and other sundry items. *See* Tub crash.

Filter   An electric network that alters the response of a certain frequency.

Filter ring   *See* Phone rings.

Flagging   A piece of paper inserted in a reel of audio tape to roughly mark a particular section of tape. Only used on tape machines not equipped with counters.

Flanging   Playing back two identical sound tracks at varying intervals for an effect.

Floor   The studio in general.

Floor plan   A scaled-to-studio drawing done by the art director to show the location of various pieces of scenery, rooms, furniture, and props. (Much of which is rearranged, either in dry rehearsal or camera blocking.)

Flutter   Discernible high-frequency distortion caused by very rapid variations in tape speed.

Fly the rag   In variety show jargon, "raise the curtain!"

Foldback   A system for getting sound information to the floor, either by headset or individual speakers.

Foleying   A sound effects technique named for Jack Foley, a film sound effects editor. What makes this technique unique in films is that the effects are laid in "manually" and not cut in with film. A Foley stage has all the props and resemblance of an old live radio studio.

Frequency   The number of cycles completed in one second.

Frequency response   The measurements of an audio system to faithfully reproduce a sound signal.

Gain   Output audio signal level.

Goosing levels   A quick acceleration of levels for brief periods of time.

Grabbing shots   A term that indicates the camera people are ad libbing camera shots to the control room for the director's approval. Widespread on remotes or other unrehearsed situations.

Graphic equalizer   A piece of equipment that allows you, with the use of narrow bandwidth faders, to see the frequency response you have chosen.

Hard wire   Connecting various pieces of equipment with wire.

Harmonics   Frequencies that are whole-numbered multiples of the fundamental.

Headroom   (1) The amount of spread you allow between the average sound levels and the loudest anticipated level. (2) The amount of space between the performer and the overhead boom microphone.

Hertz   Sound wave cycles per second.

High end   The treble portion of the frequency spectrum.

Hokey   The word used to describe extremely broad or low comedy; derived from Hokum, a vaudeville comedian. Interchangeable with "corny" or "Mickey Mouse."

Hubbub   The term used for background voices. Aptly named because atmosphere people are not allowed to say audible intelligible words. By "mouthing" hubbub, they accomplish the task of lip movement and stay within the confines of the AFTRA contract. *See* Walla walla.

Impedance   Resistance to signal. Most often you hear this term as it applies to a "mismatch" or "matching" impedance.

Input   The signal coming into a piece of equipment.

In the ball park   In the general area of acceptance.

In the clear   Isolated.

In the mud   A very low sound level.

In the red   An extremely high sound level.

Inverted ring   *See* Phone rings.

Jack   An electrical connector used to patch various sound sources into a common console.

Laser   *L*ight *a*mplification by *s*timulated *e*mission of *r*adiation.

Later reflections   Those frequencies that bounce off of surfaces and cause reverberation.

Lavalier microphone   A microphone worn around the neck, affixed to a cord.

Layering   Quite literally, adding sounds to attain a certain effect.

Laying in   Inserting sounds.

Lead-in   Usually refers to the line of dialogue given at the start.

Leaking   Low-level unwanted sounds.

Line   The final outgoing signal from all the various sources.

Lips   Mouth movement seen on video, without audio. Normally happens at the beginning of a scene when the performer speaks before the audio person has opened the mike after countdown.

Live Not prerecorded.

Logo A picture or writing that identifies a company. CBS's eye; NBC's peacock.

Long Too much show, too little time. Either the pace must be picked up, or cuts need to be made.

Loop A tape spliced in such a manner that it runs continuously without a beginning or an ending.

Looping A technique developed for theatrical films. Clips of film that were either shot silent or need to be redubbed are spliced together so that they form a "loop." These clips are projected onto a screen, and the actors and actresses say their dialogue in sync with the mouth movement on the screen.

Losing the light A film term used on location when the sun is going down. In television, unless at an outdoor location, it loosely means "it's getting late."

Loudness Inasmuch as we all hear the amplitudes of sounds in accordance with our own hearing ability, this is the one word that has caused more arguments than any other word in audio . . . certainly in sound effects.

Loudspeaker A transducer that converts electrical energy into acoustical energy. Perhaps the most common loudspeaker is found in the telephone.

Magnetic tape A plastic base material coated with fine particles of ferric oxide (iron particles).

Magnitude *See* Amplitude.

Masking Covering one sound with another of greater amplitude.

Master pot A *pot*entiometer that controls the sound output of an audio console.

Microphone A transducer that converts sound pressure waves to an electrical signal.

Midi *M*usical *i*ntrument *d*igital *i*nterface.

Midrange Those frequencies that fall into the 200 Hz to 2,000 Hz frequency range. These frequencies are the ones that are most often used in our daily conversations.

Mix The term used to describe taking various sound sources and putting them together in such a manner so that they are all in the proper perspective.

MOS An acronym for the film term: *M*it *O*ut *S*ound. Some of the early directors in Hollywood, whose first language was German, had trouble with the word *with*. So instead of saying, "We are recording this scene without sound," the words came out, "We are recording this scene *mit*out sound." Because of this word corruption, *MOS* literally means "without sound."

Muddy When an audio level is too low, it is said to be in the mud or muddy. Conversely, when an audio level is too loud, it is said to be in the red.

Needle drop Refers to a special fee paid to a music publisher for the privledge of using their copyrighted song on the air. The practice dates back to the days when all music was on discs. Each time a song is played is a "needle drop" and requires a payment to the publisher. Although this can become very expensive, playing songs without the permission of the publisher is a great deal more expensive if you are caught.

Noise Excluding distortion, any spurious electrical disturbance.

Notch filter Frequency attenuator used to tune out a very narrow band of frequencies.

Octave Octaves are perceived as equal pitch intervals that have a frequency ratio of 2 : 1.

Off mike Away from the axis of the microphone.

Omnidirectional mike A microphone with a 360-degree response pattern.

On mike Signals directed on axis; or the most responsive part of the microphone.

Onomatopoeia The naming of an object or action by a vocalization of the sound it produces. Examples include zing, buzz, hiss, and boing.

Oscillator A device that generates tones.

Overdubbing Playing back a track of information while laying down new material; the sum of all the different tracks is recorded as the finished "overdubbed" track.

Overlap When one sound comes in over another, without any pause in between. When it occurs between two actors and there is a mistake, it is impossible to edit cleanly and must be taken back to the first clean opening.

Overloading Feeding a signal too loud for corresponding equipment to handle.

Pad (1) An attenuator placed in a system to reduce sound levels. (2) A production term meaning a time allowance.

Pan Following the action of a subject by moving only the camera.

Panto Short for pantomiming.

Payolla Whereas the *plug* is an accepted business procedure conducted in broadcasting, *payolla* is more for personal gain and is discouraged.

Pedestal The mechanism on the camera that allows it to move up and down. Most often simply referred to as "ped" up or down.

Perspective The sound heard in relationship to the distance of the origin. It always is in relationship to the principal listener. In radio, it was the person on mike, in television, it is the primary person on camera.

Phon A unit of loudness level related to the ears' subjective impression of signal strength.

Phone rings There has always been some confusion as to what is the correct term to use when cueing the sound effects phone rings you hear when you place a call to someone else. The three most applicable terms are *Filter ring, inverted ring,* and *line ring.* Of the three, *inverted ring* is used most often.

Pickup Going back and redoing lines or business.

PL Private *l*ine.

Playoff Musical ending.

Plug To mention or show a product, either for money or for some other consideration.

Post production The room where the final show is put together for air.

POV Point *of* view.

Prerecord A recording done prior to a taping or broadcast.

Presence (1) A sound that is perceived as being bright and natural. (2) Nonspecific sounds used to smooth out edit points and give a feeling of life to a sound-deadened studio. "Presence" frequencies are on the upper end, somewhere between 2,000 and 8,000 Hz.

Preset monitor The monitor that reveals what picture is going to be taken next.

Preview monitor Same as preset monitor.

Pull up Shortening the space between two pieces of information, regarding either audio or video.

Punch up (1) When the technical director cuts from camera to camera he pushes appropriate buttons. From this action we get "punch up"— to put a particular camera's picture on the line or air monitor. (2) On a comedy show, making something funnier. (3) In general, any situation that needs more energy or emphasis.

Putting it on its feet Actually walking through or rehearsing a scene.

Ready A word given in preparation for the actual cue.

Remote Originating from a place other than the studio.

Repo Reposition.

Resonance The condition when the applied frequency is equal to the normal frequency of vibration of the system.

Reverberation Repeated sound reflections after the original sound is out.

Riding gain   Manually moving faders to keep the audio or video levels within acceptable broadcasting standards.

Rocking   Moving a short section of tape across the reproduction head by manually turning the tape reels. *Rocking* is of special value in finding an edit point.

Room noise   *See* Presence.

Round robin   Originating from one source and going to other locations before returning to the original source.

Run through   Same as rehearsal.

SA   Studio Announce. *See* Talkback.

Sampling   When an analog signal is sent through a digital converter, fragments of the signal's amplitude are changed into binary digits that are representative of the signal. The greater the number of digits, the more accurately the analog signal will be "seen" by the digital computer.

Segue   A commonly misused term, *segueing* from one tape or record to another always happens at the *conclusion* without interruption or talk. Not to be confused with *Crossfade*.

Selling   Ad libbing during rehearsal for the express purpose of having an idea heard or seen in hopes of it being incorporated into the show.

Sel sync   A trademark of the Ampex Corporation, it is short for "selective synchronization." This process enables you to record and play back at the same time in perfect sync. This is accomplished by using the record head as a playback head. If you are using a four-track tape recorder, you may play back a piano tape on track no. 1, a guitar tape on track no. 2, drums on track no. 3, and then record your voice on track no. 4 . . . without any delay and without the danger of erasing the other three tracks. You will have one composite track, with the voice accompanying the music tracks in perfect sync.

Separation   The distance between two performers.

SFX   Sound effects. Originally a film term.

Short   Too much time and not enough show. *See* Long.

Shtick   Action or physical business; usually refers to comedy.

Sibilance   A "hissing" sound caused by hitting S's, C's, and similar sounds too hard or too close to the mike.

Signal to noise   The ratio of the signal voltage to the noise voltage.

Slating   An audio and picture announcement of the name of the show, the show number, month, day, year, director, and take number. After the initial slating, additional pickups are simply referred to as "take" numbers.

Slipping   After the turntable has come up to proper speed, the operator gently applies pressure to the record, holding it in one place while allowing the turntable plate beneath it to turn freely. When the cue comes, the operator releases the record and opens the fader. *See* Torque.

SMPTE time code The standard time code for editing purposes. The acronym for *Society of Motion Picture and Television Engineers.*

SOF *Sound on film.*

Sound locker Another name for a television sound effects room.

Source The origin of a sound.

Splice An edit point.

Squeeze Camera direction to tighten a camera shot.

Stacked Bad positioning of actors so that they are standing in a row.

Stand by The command given in preparation of a cue.

Sting Sharp tremolo of music used for emphasis or suspense.

Stretch Slow the pace of the dialogue and action for time purposes.

Stylus The needle of a pickup arm on a turntable. Now usually diamond tipped; early ones were made of steel and even bamboo!

Sustain Hold a musical note or sound for a certain duration.

Sustaining Without commercial sponsorship.

Sweetening Adding elements to a program after editing.

Switcher A system of filters built into the audio console and patchable to the cameras.

Sync Syncronizing. The term used for following action as closely as possible.

Sync pulse The tone that syncs up the audio with the picture in a two-system recording.

Tail The natural decay or ending of a sound.

Tails out The tape has been rewound so that the end of the tape is now at the beginning.

Takeup reel The reel on which tape winds after it leaves the head assembly.

Talkback *see* SA.

Tally lights The small lights on a camera that, when lit, indicate the camera is on.

Tape loop A tape spliced to run continuously.

TC Tentative cut. On soap operas, when the writers sense that their scripts are going to run long, they block out TCs rather than have the director or producer try to find cuts that might interfere with future story lines.

Telegraph To indicate by word or action that something is about to happen.

Timbre The individual color or quality of a sound.

Torque The efficiency of a tape machine or turntable motor to get up to the proper speed in the least amount of time. Machines that lack good torque cause the audio at the beginning of a tape or record to start too slow and as a result, the opening audio is uneven and indistinct. When this happens, the tape or record is said to "wow." *See* slipping.

Transducer An instrument that converts one form of energy to another.

Tub Crash A manual crash involving a large laundry tub filled with pots, pans, cutlery, and whatever else you can think of that might sound funny.

Under five An actor who has five or fewer speaking lines.

Underscore Providing music or sound effects to enhance a scene. Almost exclusively applies to music. *See* Laying in.

Upcut Audio loss of part of a word. *See* Lips.

Upstage Away from the footlights or camera.

Upstaging Applies to the action or business a performer does that attracts attention to himself while his fellow performer is supposed to be the center of attention.

Variable speed control A device used to alter the speed of a tape machine or turntable.

Volume control Same as fader or potentiometer.

Volume meter Often referred to as a VU meter. Measures the current flow of an acoustical signal.

Walla walla The nonsense term in films for background voices. *See* Hubbub.

Winging Same as ad libbing.

Wireless microphone A small microphone connected to a battery-powered transmitter capable of sending a signal several hundred yards to the mixing console.

Wrap The end; to conclude; to finish.

BIBLIOGRAPHY

Alten, Stanley R. *Audio in Media*. Belmont, Calif.: Wadsworth Publishing, 1981.

Blake, Larry. *Film Sound Today*. Hollywood, Calif.: Reveille Press, 1984.

Bronfeld, Stewart. *How to Produce a Film*. Englewood Cliffs, N.J.: Prentice-Hall, 1984.

Buxton, Frank, and Bill Owen. *The Big Broadcast*. New York: Avon Books, 1973.

Clifford, Martin. *Microphones*. 2d ed. Blue Ridge Summit, Pa.: Tab Books, 1982.

Encyclopedia Americana, Electronics ed, s.v. Copyrighted Grolier, Inc., 1964.

Evans, Alvis J., Jerry D. Mullen, and Danny H. Smith. *Basic Electronic Technology*. Dallas: Texas Instruments Information Publishing Center, 1985.

Horn, Delton E. *Digital Electronic Synthesizers*. Blue Ridge Summit, Pa.: Tab Books, 1980.

McGill, Earle. *Radio Directing*. New York: McGraw-Hill, 1940.

Smith, F. Leslie. *Perspectives on Radio and Television: Telecommunications in the United States*. New York: Harper & Row, 1986.

Walker, Alexander. *The Shattered Silents*. New York: William Morrow & Co., 1979.

Weis, Elisabeth, and John Belton. *Film Sound: Theory and Practice*. New York: Columbia University Press, 1985.

White, Glenn D. *The Audio Dictionary.* Seattle, University of Washington Press, 1987.

Woram, John M. *The Recording Studio Handbook.* Plainview, N.Y.: Elar Publishing, 1985.

Krows, Arthur Edwin. *Krows Equipment for Stage Production. A Manual For Scene Building.* New York: D. Appleton, 1928.

INDEX